Studies in Computational Intelligence

Volume 1216

The series "Studies in Computational Intelligence" (SCI) publishes new developments and advances in the various areas of computational intelligence—quickly and with a high quality. The intent is to cover the theory, applications, and design methods of computational intelligence, as embedded in the fields of engineering, computer science, physics and life sciences, as well as the methodologies behind them. The series contains monographs, lecture notes and edited volumes in computational intelligence spanning the areas of neural networks, connectionist systems, genetic algorithms, evolutionary computation, artificial intelligence, cellular automata, self-organizing systems, soft computing, fuzzy systems, and hybrid intelligent systems. Of particular value to both the contributors and the readership are the short publication timeframe and the world-wide distribution, which enable both wide and rapid dissemination of research output.

Indexed by SCOPUS, DBLP, WTI AG (Switzerland), zbMATH, SCImago.

All books published in the series are submitted for consideration in Web of Science.

Muhammad Firoz Mridha · Nilanjan Dey
Editors

Federated Learning in Health Care Technology

Springer

Editors
Muhammad Firoz Mridha
Department of Computer Science
American International
University-Bangladesh
Dhaka, Bangladesh

Nilanjan Dey
Department of Computer Science
and Engineering
Techno International New Town
Kolkata, West Bengal, India

ISSN 1860-949X ISSN 1860-9503 (electronic)
Studies in Computational Intelligence
ISBN 978-981-96-8352-9 ISBN 978-981-96-8353-6 (eBook)
https://doi.org/10.1007/978-981-96-8353-6

This Springer imprint is published by the registered company Springer Nature Singapore Pte Ltd.
The registered company address is: 152 Beach Road, #21-01/04 Gateway East, Singapore 189721, Singapore

Preface

The rapid advancements in healthcare technology are paving the way for unprecedented innovations, and *Federated Learning in Health Care Technology* stands at the forefront of this revolution. This book offers an in-depth exploration of federated learning (FL), a groundbreaking approach that facilitates collaborative data analysis while ensuring patient privacy and data security. As healthcare systems worldwide generate vast amounts of data, the challenge lies in harnessing this information without compromising confidentiality. Federated learning addresses this by allowing multiple institutions to collaborate on machine learning models without sharing sensitive data.

In this context, we focused on the foundational principles of FL, illustrating how it enables the aggregation of decentralized data to improve diagnostic accuracy, develop personalized treatment plans, and enhance overall healthcare outcomes. We present real-world applications across various medical fields, from predictive analytics in chronic disease management to precision medicine and beyond. Additionally, we discuss the ethical and regulatory landscapes, providing insights into the challenges and solutions associated with implementing FL in healthcare.

This book is designed for a diverse audience, including researchers, healthcare practitioners, data scientists, and policymakers. It aims to bridge the gap between cutting-edge technology and practical medical applications, offering a comprehensive guide to leveraging FL for healthcare innovation. By integrating case studies, technical details, and ethical considerations, we provide a holistic view of how FL can revolutionize the healthcare industry.

Ultimately, *Federated Learning in Health Care Technology* aspires to inspire and equip readers with the knowledge to drive forward the next wave of data-driven healthcare solutions, fostering a future where technology and medicine work hand in hand to deliver superior patient care.

This book reflects a rich diversity of contributions from experts across the globe, highlighting the collaborative efforts of researchers from countries including Bangladesh, the US, Canada, Japan, Australia, and Saudi Arabia. The chapters explore a wide range of topics, as outlined below:

- Chapter "Federated Learning in Healthcare Technology: Challenges, Solutions and Opportunities" by Abha et al. sets the stage by discussing the potential of FL in healthcare and its challenges and opportunities.
- Chapter "Decentralized Tumor Classification with Federated Learning: A Privacy-Preserving Approach" by Mir et al. focuses on FL's application in tumor classification while maintaining data privacy.
- Chapter "Transforming Healthcare Analytics: Federated Learning for Privacy-Preserving Liver Lesion Detection and Classification" by Islam et al. discusses FL's role in improving liver lesion detection.
- Chapter "Privacy-Preserving Kidney Stone Detection from X-Ray Image using Federated Learning" by Plabon et al. highlights FL's capability to securely detect kidney stones.
- Chapter "Federated Learning for Privacy-Preserving Healthcare Analytics: A Novel Framework for Fraud Detection in Healthcare" by Rafi et al. introduces an FL framework to combat fraud in healthcare.
- Chapter "Federated Learning Strategies for Confidential Leukemia Detection from Medical Images" by Plabon et al. focuses on FL applications for leukemia detection.
- Chapter "Secure Collaborative Learning for CSV-Based Ovarian Cancer Diagnosis: A Federated Approach" by Diba et al. describes an FL approach for ovarian cancer diagnosis.
- Chapter "Privacy-Preserving Rheumatism Detection Using Federated Learning" by Diba et al. showcases FL's utility in chronic disease management.
- Chapter "Enhancing Gall Bladder Cancer Detection by Privacy-Preserving Federated Learning: A Comprehensive CSV-Based Classification Approach" by Diba et al. examines the potential of FL in gall bladder cancer detection.
- Chapter "Cataract Detection Using Federated Learning: A Decentralized Approach for Cross-Institutional Collaboration of Data Parameters" by Shourov et al. presents FL's application in cataract detection.
- Chapter "Advancements and Challenges in Federated Learning for Privacy-Preserving Smart Healthcare: A Review" by Rimi et al. reviews the advancements and challenges in FL for healthcare.

The editors extend their gratitude to the exceptional authors and referees for their valuable contributions to this book. Their dedication and expertise have been instrumental in creating this comprehensive resource. Special thanks are also due to the Springer team for their unwavering support.

Dhaka, Bangladesh — Muhammad Firoz Mridha
Kolkata, India — Nilanjan Dey

Contents

About the Editors

Muhammad Firoz Mridha (Senior Member IEEE, Professional Member ACM) is currently working as an Professor in the Department of Computer Science, American International University-Bangladesh (AIUB). He also worked as Associate Professor and Chairman in the Department of Computer Science and Engineering, Bangladesh University of Business and Technology (BUBT) from 2019 to 2022 and as a CSE department faculty member at the University of Asia Pacific and as a graduate head from 2012 to 2019. He is the founder and director of Advanced Machine Intelligence Research Lab (AMIR Lab). He received his Ph.D. in the domain of AI from Jahangirnagar University in the year 2017. For more than 15 (Fifteen) years, he has been with the master's and undergraduate students as a supervisor of their thesis work. He has **authored/edited** several books with **Springer** and published more than 200 Journal and Conference papers. His research work contributed to the reputed Journal of Scientific Reports Nature, Knowledge-Based Systems, Artificial Intelligence Review, Engineering Applications of Artificial Intelligence, IEEE Access, Sensors, Cancers, Biology and Applied Sciences etc. His research interests include Computer Vision (CV), Artificial Intelligence (AI), Machine Learning (ML), Deep Learning, and Natural Language Processing (NLP) etc. He has served as a program committee member in several international conferences/workshops. He served as an Editorial Board Member of several journals including **PLOS ONE** Journal. He has served as a reviewer of reputed journals like IEEE Transactions on Neural Networks,

Artificial Intelligence Review, IEEE Access, Knowledge base System, Expert System, Bioinformatics, Springer Nature etc. He was included among the top 2% of scientists worldwide in 2024 edition of the Stanford University/Elsevier.

Nilanjan Dey (Senior Member, IEEE) received the B.Tech., M.Tech. in information technology from West Bengal Board of Technical University and Ph.D. degrees in electronics and telecommunication engineering from Jadavpur University, Kolkata, India, in 2005, 2011, and 2015, respectively.

Currently, he is Associate Professor with the Techno International New Town, Kolkata and a visiting fellow of the University of Reading, UK. He has authored over 300 research articles in peer-reviewed journals and international conferences and 40 authored books. His research interests include medical imaging and machine learning. Moreover, he actively participates in program and organizing committees for prestigious international conferences, including World Conference on Smart Trends in Systems Security and Sustainability (WorldS4), International Congress on Information and Communication Technology (ICICT), International Conference on Information and Communications Technology for Sustainable Development (ICT4SD) etc.

He is also the Editor-in-Chief of International Journal of Ambient Computing and Intelligence, Associate Editor of IEEE Transactions on Technology and Society and series Co-Editor of Springer Tracts in Nature-Inspired Computing and Data-Intensive Research from Springer Nature and Advances in Ubiquitous Sensing Applications for Healthcare from Elsevier etc. Furthermore, an Editorial Board Member of Complex and Intelligence Systems, Springer, Applied Soft Computing, Elsevier and he is an International Journal of Information Technology, Springer, International Journal of Information and Decision Sciences etc.

He is a Fellow of IETE and member of IE, ISOC etc.

Federated Learning in Healthcare Technology: Challenges, Solutions and Opportunities

Esm E Moula Chowdhury Abha, Md. Imtiaj Alam Sajin, Md Wahiduzzaman Suva, Shanjida Akter, Mushfiqur Rahman Abir, Md. Asgor Hossain Reaj, and Jahir Ibna Rafiq

Abstract Federated Learning (FL) offers a novel method for training models collaboratively across multiple distributed data sources without the need to collect data in a central location. For healthcare organizations, where data protection and security are of utmost importance, FL offers a promising solution for utilizing insights from diverse datasets. At the same time, it ensures the preservation of patient confidentiality. This paper examines the difficulties and recent developments in applying FL to smart healthcare. We propose a framework that addresses technical and non-technical challenges, by utilizing privacy-preserving techniques, personalized learning algorithms, efficient communication protocols, and infrastructure support. The framework, based on existing literature and practical implementations, provides a complete roadmap for implementing FL in healthcare service while safeguarding patient privacy.

Keywords Decentralized learning · Federated learning · Privacy-preserving · Healthcare

E. E. M. C. Abha · Md. I. A. Sajin · M. W. Suva · M. R. Abir · Md. A. H. Reaj
Department of Computer Science, American International University-Bangladesh, Dhaka, Bangladesh

S. Akter
Department of Computer Science and Engineering, North South University, Dhaka, Bangladesh
e-mail: shanjida.akter01@northsouth.edu

J. I. Rafiq (✉)
Biomedical Engineering Department, University of Toronto, Toronto, Canada
e-mail: JahirIbna.Rafiq@uhn.ca

M. F. Mridha and N. Dey (eds.), *Federated Learning in Health Care Technology*, Studies in Computational Intelligence 1216, https://doi.org/10.1007/978-981-96-8353-6_1

1 Introduction

Usage of AI Techniques are getting more popular on healthcare day by day. Specifically, machine learning (ML) and deep learning (DL) have shown success in many sectors, including healthcare [1]. However, challenges such as training limitations and concerns about data security have persisted in the healthcare field. Federated Learning (FL) offers a promising approach to addressing these challenges effectively.

Healthcare systems generate a vast amount of data regularly. This includes patient records, medical images, and telemetry from wearable devices. To utilize this data for collaborative work between institutions is essential for advancing patient care and medical research. However, traditional methods of data exchange face several challenges such as privacy concerns, regulatory constraints, and logistical complexities. Federated Learning (FL) has emerged as a promising solution, as it enables model training to occur locally on client devices or servers, with only aggregated updates sent to a central server for integration.

In the field of healthcare informatics, Federated Learning (FL) has shown significant advancements in recent years. It has tackled many challenges with innovative and unique approaches. Privacy-preserving techniques such as differential privacy and secure aggregation have played a vital role [2, 3]. These methods ensure that individual patient information remains confidential. Which also allows parallelly training collaborative model across distributed datasets and update the center model. Moreover, studies have shown the potential of FL in specific healthcare applications, such as cancer detection, heart disease prediction, and sleep quality prediction. These demonstrates its capability and efficiency [4–6].

The integration of FL with advanced technologies such as blockchain and edge computing improves its security and efficiency in managing healthcare data. For example, blockchain guarantees the integrity of data and secure communication, which is essential for preserving trust in FL systems [2]. Furthermore, specialized algorithms for the healthcare industry, such as the Differential Privacy Federated Learning Healthcare IoT (DPFL-HIoT) model, enhance model parameters using local data attributes. This leads to improve performance and enhance the primacy also [7].

Despite all these, security remains a concerning factor in FL deployments. Threats such as adversarial attacks and malicious clients require strong defenses, including adversarial training and secure multi-party computation (SMC) [2, 7]. Moreover, the scalability and interoperability of FL systems require infrastructures to support diverse healthcare environments.

This paper aims to provide a detailed overview of recent developments and obstacles in FL for smart healthcare systems. After synthesizing insights from various studies, we propose a framework that integrates technical advancements with regulatory and organizational strategies. This framework not only addresses technical complexities but also ensures aligns with privacy regulations and ethical standards.

2 Recent Advancements in Federated Learning

Federated learning (FL) has gained significant attention in the smart healthcare domain due to its ability to enable collaborative learning without compromising patient data privacy. Recent advancements have made FL more useful and efficient in healthcare.

Federated learning has evolved rapidly, with improvements in algorithmic efficiency, privacy preservation techniques, and real-world implementations. One of the notable advancements is the development of privacy-preserving techniques that ensure sensitive health data remains secure while utilized for training machine learning models. Techniques such as differential privacy and secure multi-party computation have been integrated into federated learning frameworks to enhance data security [8].

Xu et al. introduced federated learning (FL) as a new approach in the field of smart health, where models can be collaboratively trained while ensuring the security and privacy of each patient's data. Recent FL technologies have focused on improving privacy-preserving techniques, enhancing communication efficiency, and supporting model personalization to meet varying needs. However, there is still an outstanding challenge in the management of data heterogeneity across institutions, ensuring strong privacy and security against advanced attacks, and managing the complexities in the regulatory landscape, for instance, in HIPAA compliance. These challenges need to be addressed to integrate FL successfully within healthcare, which calls for further research in developing standardized data protocols, advanced cryptographic methods and more resource-efficient FL systems [9].

Several frameworks have been proposed to implement federated learning in healthcare effectively. For instance, Google's TensorFlow Federated (TFF) is an open-source framework that facilitates the deployment of FL across various devices, enabling collaborative model training without centralizing the data [10]. Another significant contribution is the Federated-Averaging (FedAvg) algorithm, also developed by Google Research. FedAvg is a cornerstone of federated learning that allows for efficient and scalable training of deep neural networks on distributed data. It works by averaging the locally trained models from participating devices to produce a global model, thereby significantly improving the efficiency of federated learning and making it feasible to train complex models on decentralized healthcare data [11].

Additionally, PySyft, developed by OpenMined, is an open-source library that integrates with popular machine learning frameworks like PyTorch and TensorFlow to enable privacy-preserving federated learning. PySyft provides tools for secure multi-party computation, differential privacy, and encrypted model training, facilitating the development of privacy-preserving federated learning models and enhancing data security and compliance with healthcare regulations [12].

NVIDIA's Clara Federated Learning [13] is another noteworthy framework. Clara is a healthcare application framework that includes tools for developing federated learning applications in medical imaging and other healthcare domains. The

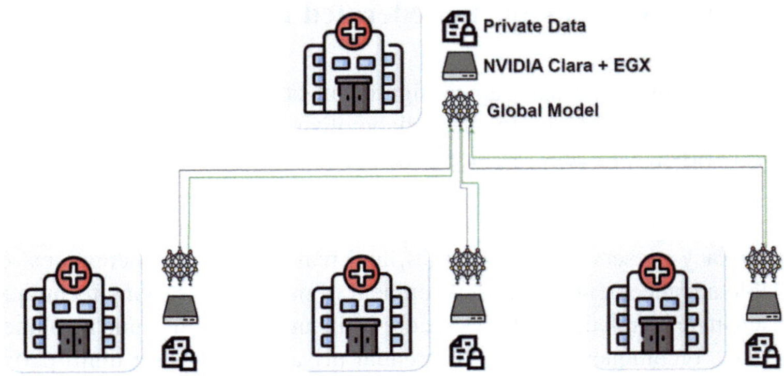

Fig. 1 Federated learning powered by NVIDIA clara

Fig. 1 represents the framework of NVIDIA Clara. The framework has secure communication protocols and supports various privacy-preserving techniques. It helps to collaborate between medical institutions to train models on sensitive medical data without sharing it. This framework has been used to improve diagnostic accuracy and patient outcomes in collaborative healthcare settings.

A specific example of FL in action is the work by Sheller et al. [14], who demonstrated the use of FL to train a neural network for brain tumor segmentation using data from multiple medical institutions. This study highlighted the potential of FL to achieve high model performance while maintaining data privacy across different institutions.

One prominent case study is the implementation of federated learning for predictive modeling in chronic disease management. Li et al. [15] explored the use of FL to develop predictive models for diabetes management by aggregating data from various healthcare providers. The study showed that FL could produce accurate predictive models comparable to those trained on centralized data, thereby validating the effectiveness of FL in real-world healthcare applications.

Another example is the collaboration between major hospitals and research institutions to develop FL models for detection [16] for prognosis. *Xu et al.* have used federated learning (FL) to build strong models that can predict how severe COVID-19 will be and outcomes of the patient. They do this using data from different places without sharing it directly [17].

Lastly, the use of secure aggregation protocols in federated learning, developed by various research institutions, ensures that individual updates from participating clients are encrypted and aggregated securely. This prevents any single client or server from accessing the raw data, enhancing privacy and security in federated learning systems. Secure aggregation has strengthened the privacy guarantees of federated learning, making it more suitable for sensitive healthcare applications [18].

3 Challenges

Federated Learning (FL) in smart healthcare has shown great promise, offering many benefits such as enhanced data privacy and improved collaborative learning across institutions. However, the implementation of FL in this field is not without its challenges. These challenges are crucial and need to be addressed to ensure the successful adoption and operation of FL in healthcare settings. Here, we present a range of issues considering both the technical and non-technical aspects [19].

3.1 Technical Challenges

One of the primary technical challenges in federated learning is ensuring **data privacy and security**. The high risk of data leakage during communication and model updates arise the concern of privacy-protection strategies. Highlighting these concerns, *Shokri and Shmatikov* underscore the profound privacy issues associated with deep learning applications. Their study emphasizes how large-scale data collection, essential for training deep learning models effectively, raises significant privacy concerns, particularly regarding sensitive user data such as images and clinical records. Although FL aims to protect privacy by keeping data local, there is still a risk of data leakage during communication and model updates. To minimize these risks requires strong implementation of advanced privacy-protection strategies [9].

Another critical issue is **Data heterogeneity**. When data is collected from different sources, variations arise due to differences in treatment methods, equipment, patient populations, medical data types, and geographical areas across institutions. These variations can significantly impact the performance of global models. Specifically, non-IID (Independent and Identically Distributed) data, which does not follow the same statistical patterns, poses challenges. There is still some scope to explore methods to handle non-IID data in federated learning [20].

Communication efficiency presents a significant challenge in federated learning (FL), where frequent exchanges between central servers and client devices can lead to high latency and bandwidth consumption [21]. To address this, optimizations in communication protocols are essential to reduce the frequency and size of updates to enhance scalability and efficiency, particularly in healthcare applications. A control algorithm that can effectively balances local model updates with global parameter aggregation, ultimately optimizing the loss function within resource constraints would be helpful in this context.

Scalability is another major challenge in federated learning. FL handles a significant number of devices that participate in the training procedure. To effectively manage these devices with limited resources, it requires complex and advanced orchestration and scheduling algorithms. This complexity is further increased by the need to maintain flawless communication between various devices and balance computational tasks [22].

3.2 *Non-technical Challenges*

Other than technological difficulties, In the healthcare sector, federated learning faces a number of non-technical difficulties which causes to raise ethical and regulatory concerns frequently. Healthcare data is heavily regulated, with laws like the Health Insurance Portability and Accountability Act (HIPAA) and the General Data Protection Regulation (GDPR) setting strict standards. Following these guidelines during the implementation of federated learning can be challenging [23].

Interpersonal trust is another important non-technical factor. Collaboration between several stakeholders, including as hospitals, research institutes, and technology suppliers, is necessary for successful federated learning. Collaboration and the sharing of data between participants depend on establishing and preserving trust. To build trust, systems are required to guarantee accountability, justice, and transparency [24, 25].

Federated learning systems involve significant computational resources and investments. Smaller healthcare providers might struggle with the costs associated with maintaining the necessary infrastructure and expertise. Developing cost-effective solutions and providing support for resource-limited settings are essential for broader adoption [26].

These difficulties draw the attention of the complexity of implementing federated learning in smart healthcare. To address these issues, ongoing research, interdisciplinary collaboration, and the development of innovative solutions are required to ensure that federated learning can be effectively and ethically integrated into healthcare systems.

4 Proposed Solutions and Hypothesis

Federated learning (FL) presents a number of challenges that call for a combined strategy that combines organizational and regulatory approaches. The proposed ideas and hypotheses are designed to help overcome the challenges we have identified and make the implementations of FL more practical and impactful. Below, we have outlined several solutions that take into account both the technical and non-technical issues we previously discussed.

4.1 *Technical Solutions*

In this section, we propose ideas aimed at addressing the challenges discussed earlier in federated learning (FL). Building upon the foundation laid out in the preceding section, our focus will be on integrating advanced privacy-preserving techniques [27, 28], optimizing model adaptability across heterogeneous healthcare datasets, and enhancing communication protocols. These ideas are pivotal for developing a comprehensive framework that not only safeguards data privacy and enhances model performance but also ensures scalability and efficiency in healthcare environments with limited resources.

To tackle the significant issues of privacy and security in federated learning (FL), it is essential to employ sophisticated methods like homomorphic encryption and secure multi-party computation (SMPC). These advanced techniques allow calculations on encrypted data without the need for decryption, thus protecting the privacy of sensitive information while enabling joint model training across several entities. By integrating homomorphic encryption and SMPC into FL frameworks, we can significantly enhance data security measures. This approach ensures robust protection against unauthorized access and data breaches, particularly crucial in healthcare and other privacy-sensitive domains where maintaining the privacy of patient data is paramount [29].

In the healthcare sector, characterized by diverse data sources and regulatory constraints, personalized federated learning algorithms represent a transformative solution. Techniques like federated multi-task learning and meta-learning are used in healthcare help models learn and adapt better. Besides, they handle different types of data from various healthcare providers. Figure 2 represents the MetaFL, local model aggregation protocol. By tailoring models to local data patterns while federating knowledge across institutions, these approaches enhance predictive accuracy and model generalizability. This advancement not only optimizes healthcare outcomes through comprehensive data utilization but also addresses privacy concerns by decentralizing sensitive patient data and ensuring compliance with stringent data protection regulations [30].

Efficient communication protocols play a pivotal role in optimizing federated learning performance, particularly in resource-constrained healthcare environments.

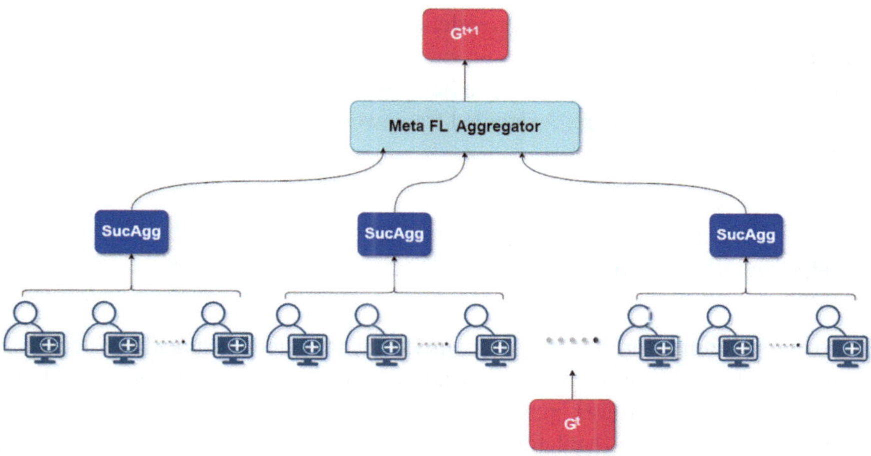

Fig. 2 Local model aggregation using MetaFL

Strategies such as quantization, which reduces the precision of model parameters without compromising accuracy, and sparsification, which focuses on transmitting only essential model updates, effectively mitigate communication overhead. Additionally, streamlined communication protocols designed for federated environments enhance scalability and reduce latency, facilitating seamless collaboration and efficient model convergence across distributed nodes. These optimizations are crucial for enabling federated learning to operate effectively in real-time healthcare applications, supporting timely data analysis and decision-making while maintaining robust system performance resources [31].

4.2 Non-technical Solutions

A strong regulatory framework is significant to ensure federated learning in healthcare complies with data protection laws like HIPAA and GDPR. This framework should provide clear guidelines for data anonymization, obtaining patient consent, and maintaining audit trails. With these clear rules, more people and stakeholders will trust federated learning and it can be used more widely [32].

As interpersonal trust between several stakeholders is important while collaboration, transparent reporting and third-party audits can make sure everyone is accountable. Blockchain technology can also help by making the whole process more transparent and secure. When everyone sees that the system is fair and reliable, they will be more willing to share data and work together. This will further lead to greater data sharing and collaborative model development [33].

Cost-effective infrastructure needed to ensure smaller healthcare providers can afford the technology needed for federated learning. For cutting costs cloud-based and solutions shared resources could be used. Besides, providing federated learning as a service (FLaaS) can make the technology accessible to all, regardless of size. This way, even small healthcare providers can join in and benefit from federated learning [34].

These suggested solutions and hypotheses tackle the key challenges in federated learning for smart healthcare. A detailed framework development approach is given in the next section.

5 Proposed Framework

The proposed framework is designed to tackle the main technical and non-technical challenges of implementing federated learning (FL) in smart healthcare. It includes advanced privacy measures, personalized learning methods, efficient communication protocols, strong regulatory compliance, trust-building strategies [35, 36], and affordable infrastructure support. The aim is to build a secure, scalable, and collaborative environment for analyzing healthcare data and training models.

Architecture Overview The proposed federated learning framework consists of three main components:

1. Client Nodes
2. Central Server
3. Communication Layer.

5.1 Client Nodes

In the proposed federated learning framework, client nodes serve as the foundational entities representing local healthcare data sources, including hospitals, clinics, and wearable devices. These nodes perform multiple critical tasks essential to the federated learning procedure. The Fig. 3 details the framework of Client Node. Initially, client nodes engage in local data processing. Healthcare data is prepared by removing inconsistencies and formatting it correctly for model training. Data anonymization methods are employed to protect patient confidentiality and adhere to regulatory requirements.

Subsequently, client nodes undertake local model training using the preprocessed data. Local machine learning models are trained using personalized federated learning algorithms designed to handle the unique and distributed characteristics of healthcare data. These algorithms effectively manage non-IID (non-Independent and Identically Distributed) data distributions, ensuring robust model training across diverse datasets [25].

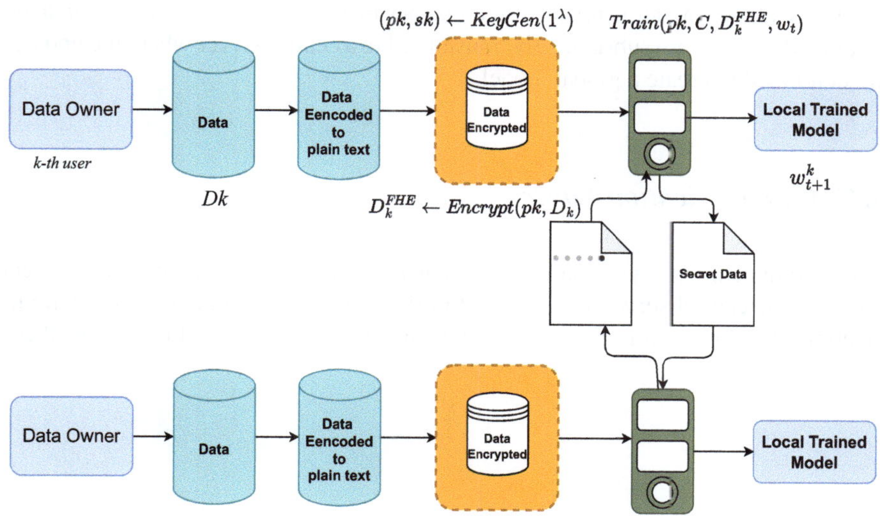

Fig. 3 Local model training using fully homomorphic encrypted data

Client nodes also integrate privacy-preserving techniques such as fully homomorphic encryption (FHE) and secure multi-party computation (SMPC) to maintain data integrity and confidentiality during local model updates. FHE allows computations to be performed on encrypted data without necessity of decryption while ensuring the privacy is maintained. To incorporate Fully Homomorphic Encryption (FHE) in the training process where for k^{th} user the encrypted data D_k is used to train the model W_{t+1}^k using the previous model W_t, we need to Generate the public and private key pair using the KeyGen algorithm.

$$(pk, sk) \leftarrow \text{KeyGen}(1^{\lambda})$$

Here, λ is the security parameter. Now we can encrypt the data D_k using the public key.

$$D_k^{\text{FHE}} \leftarrow \text{Encrypt}(pk, D_k)$$

Now, the Train algorithm can be used to perform training on the encrypted data. This involves the current model w_t and the encrypted data D_k^{FHE} to update the model to local model w_{t+1}^k. Let C be the circuit representing the training function.

$$w_{t+1}^k \leftarrow \text{Train}(pk, C, D_k^{\text{FHE}}, w_t)$$

Here, each client trains global model w_t that they received on its local dataset D_k. The training is performed over several local epochs. The client's local data D_k is split into batches of size B. And for each batch b, η is the learning rate and $\ell(w; b)$ is the loss function evaluated on batch b.

This equation ensures that the data D_k are encrypted, processed while encrypted, and the resulting model is updated in a manner that preserves privacy. The updated local models W_{t+1}^k are subsequently transmitted to a central server where the updates are combined to create a global model.

5.2 Communication Layer

The communication layer enables the transmission of model updates from client nodes to the central server, ensuring both privacy protection and effective communication. As illustrated in Fig. 4, stochastic quantization is utilized to achieve these objectives.

Algorithm 1 Stochastic Quantization and Encryption [1]

Require: N local devices, each associated with a local dataset $D_k \in \mathcal{D}$ ($k = 1, \ldots, N$); stochastic quantizer parameters (n, C_q)

for $t = 0, \ldots, T-1$ **do**

At the start of each round, the server broadcasts the global weight ω_t to n randomly selected local devices from the total pool of N devices.

for each local device k in parallel **do**

The local weight w_t is updated to w_t^k using local computations.

The weight difference is computed as $h_t^k = w_{t+1}^k - w_t$.

To ensure the difference remains bounded, it is clipped:

$$\tilde{h}_t^k = h_t^k \cdot \min\left(1, \frac{C_q}{2\|h_t^k\|}\right).$$

Noise is added to the clipped difference to preserve privacy:

$$\bar{h}_t^k = \tilde{h}_t^k + \mathcal{N}(0, \sigma^2).$$

The noisy difference is quantized:

$$\overline{h}_t^k = Q(\bar{h}_t^k; n, C_q).$$

The quantized difference is encrypted:

$$E(\overline{h}_t^k) = \text{Encrypt}(\overline{h}_t^k, K).$$

Finally, the encrypted data $E(\overline{h}_t^k)$ is encoded and uploaded to the server.

end for

The server decrypts the uploaded data to retrieve the quantized differences:

$$\overline{h}_t^k = \text{Decrypt}(sk, E(\overline{h}_t^k)).$$

The global weight is updated based on the aggregated differences:

$$w_{t+1} \leftarrow w_t + \sum_{k=1}^{n} \alpha_k \overline{h}_t^k.$$

end for

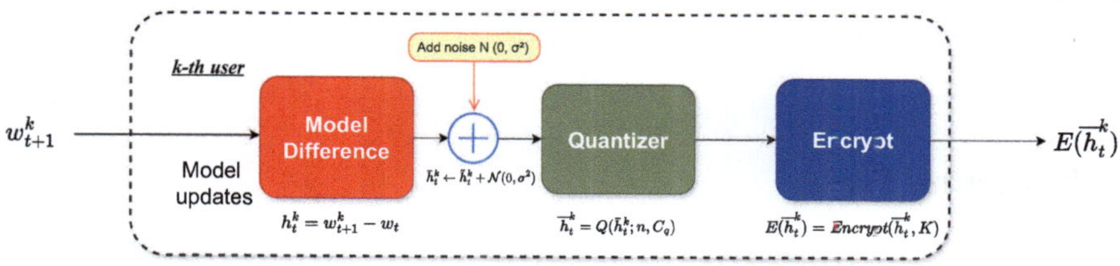

Fig. 4 Quantizing the model difference for efficient communication

Quantization Technique

We introduce an n-level stochastic quantizer with clipping, described as follows:

Definition 1 (*n-Level Stochastic Quantizer with Clipping*) Let w be an input vector with dimension d, and the clipping threshold be defined as $C_q > 0$. The input vector is clipped as

$$\tilde{w} = w \cdot \min\left\{1, \frac{C_q}{\|w\|}\right\}.$$

The codomain of the n-level quantizer is represented as S^d, where:

$$S = \{B(r) \mid B(r) = -C_q + \frac{2C_q}{n-1}r,\ r = 0, 1, \ldots, n-1\}.$$

here, an n-level stochastic quantizer with clipping threshold C_q, denoted as $Q(w; n, C_q)$, randomly maps each clipped element $\tilde{w}_k$ to one of the two adjacent quantization levels, as

$$Q(w; n, C_q)_k = \begin{cases} B(r), & \text{w.p.} \frac{B(r+1)-\tilde{w}_k}{B(r+1)-B(r)} \\ B(r+1), & \text{w.p.} \frac{\tilde{w}_k - B(r)}{B(r+1)-B(r)} \end{cases}$$

where $Q(w; n, C_q)_k$ is the k-th element of $Q(w; n, C_q)$, and $B(r) \leq \tilde{w}_k \leq B(r+1)$.

Encryption

For encryption, we use a standard encryption technique such as AES (Advanced Encryption Standard) to encrypt the quantized model updates before uploading to the server. The encryption function is represented as

$$E(\overline{h}_t^k) = \text{Encrypt}(\overline{h}_t^k, K)$$

where $\overline{h}_t^k$ is the quantized model difference and K is the encryption key.

This combined approach of quantization and encryption ensures that the communication between client nodes and the central server is both privacy-preserving and communication-efficient.

5.3 Central Server

Conversely, the central server manages the distributed learning process by organizing training tasks and gathering model updates from client nodes. In the center, model fusion is controlled by a central server using the Federated-Averaging (FedAvg) algorithm to merge local model updates into one global model, as described by McMahan et al. [37]. The workflow of FedAvg algorithm is shown in the Fig. 5. The FedAvg algorithm works as follows: The global model weights w_0 are initialized

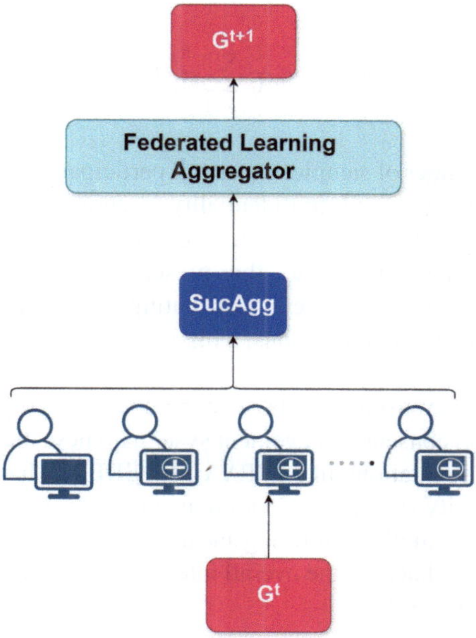

Fig. 5 FedAvg algorithm for aggregation

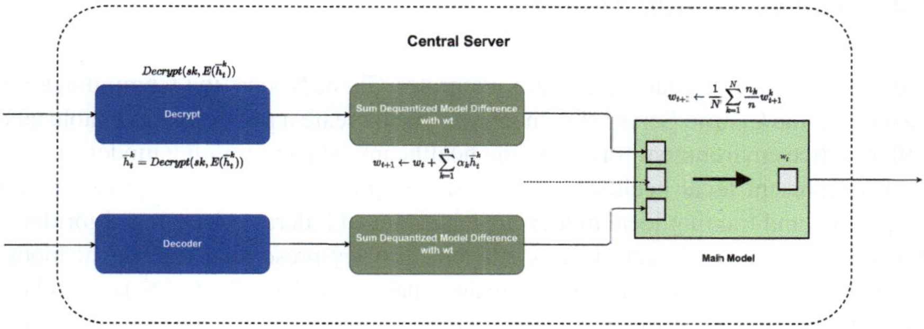

Fig. 6 Model dequantization and aggregation in central server

in Central Server, shown in Fig. 6. This initial model is then distributed to all client nodes. For each communication round $t = 1, 2, \ldots,$ the server selects a random subset S_t of $\max(C \cdot N, 1)$ clients, where C is the fraction of clients selected, and N is the total number of clients. This selection ensures that only a portion of clients participate in each round, which helps in reducing communication overhead. The recovered local model weights after the communication layer on client k are denoted as w_{t+1}^k. The central server collects updates from the local models of all selected clients. It combines these updates using a weighted average, with the weights based on the number of samples n_k each client has. This ensures that clients with more data have a greater influence on the aggregated model. The global model is updated as follows:

$$w_{t+1} \leftarrow \frac{1}{N} \sum_{k=1}^{N} \frac{n_k}{n} w_{t+1}^{k}$$

where n is the total number of samples across all participating clients, $n = \sum_{k=1}^{N} n_k$.

To ensure data integrity and confidentiality, secure aggregation protocols are employed. These protocols allow the central server to aggregate the model updates without accessing the actual local data, thus preserving client privacy. Additionally, the central server plays a crucial role in distributing the updated global model back to the clients for further training rounds. This iterative process continues until the global model converges.

The central server also ensures regulatory adherence by incorporating mechanisms such as audit trails and consent management systems. These mechanisms are crucial for compliance with regulations like HIPAA and GDPR, which mandate stringent data privacy and security standards. Continuous monitoring and supervision by the central server help in promptly identifying and addressing potential security breaches or anomalies, thereby enhancing the overall integrity and robustness of the federated learning system [11, 17, 32].

5.4 Implementation Strategy

This framework for smart healthcare integrates Client Nodes, the Communication Layer, and the Central Server shown in Fig. 7 is to create a protected, adaptable, and collaborative environment for analyzing healthcare data and training models. Client Nodes represent local healthcare data sources, processing and anonymizing data for privacy and training local models using advanced federated learning algorithms. The Communication Layer ensures efficient, privacy-preserving transfer of model updates through stochastic quantization and encryption. The Central Server orchestrates the distributed learning process, securely aggregating model updates with the Federated-Averaging algorithm and ensuring regulatory compliance with standards

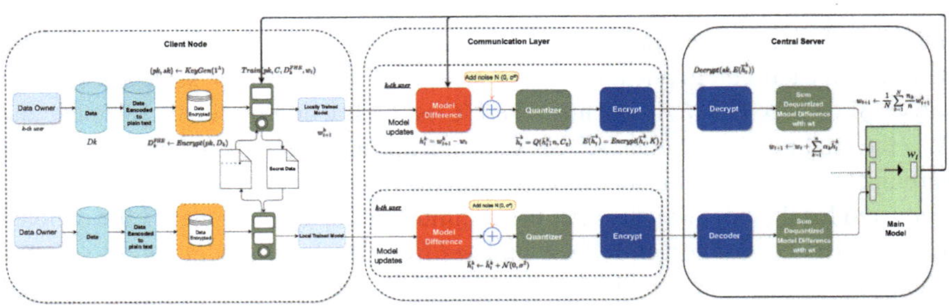

Fig. 7 Complete framework of FL in healthcare system

like HIPAA and GDPR. Complementing these technical solutions, the framework addresses non-technical challenges by establishing a strong regulatory framework, promoting transparency and accountability through third-party audits and blockchain technology, and providing cost-effective infrastructure like Federated Learning as a Service (FLaaS). This approach ensures trust and inclusivity for widespread adoption and improved healthcare outcomes.

6 Conclusion

Recent advancements in federated learning (FL) have significantly improved its efficiency and practicality in the healthcare sector, particularly through enhanced privacy-preserving techniques, personalized algorithms, and optimized communication protocols. Despite these advancements, FL faces technical challenges such as ensuring data privacy, managing data heterogeneity, and optimizing communication, along with non-technical challenges like regulatory compliance and cost-effective infrastructure. To address these, we proposed a comprehensive FL architecture comprised with client nodes, a communication layer, and a central server, each employing optimal, feasible, and privacy-preserving methods. This framework aim to create a secure, scalable, and collaborative environment for healthcare data analysis. This ultimately leads to better patient outcomes and innovative medical research advancements.

References

1. B. Shickel, P. Tighe, A. Bihorac, P. Rashidi, Deep EHR: a survey of recent advances in deep learning techniques for electronic health record (EHR) analysis (2017). https://doi.org/10.48550/ARXIV.1706.03446
2. D. Li, D. Han, T. Weng et al., Blockchain for federated learning toward secure distributed machine learning systems: a systemic survey. Soft Comput. **26**, 4423–4440 (2021). https://doi.org/10.1007/s00500-021-06496-5
3. Q. Li, Z. Wen, Z. Wu, S. Hu, N. Wang, Y. Li, X. Liu, B. He, A survey on federated learning systems: vision, hype and reality for data privacy and protection. IEEE Trans. Knowl. Data Eng. **35**(4), (2023). https://doi.org/10.1109/TKDE.2021.3124599
4. G. Long, T. Shen, Y. Tan, L. Gerrard, A. Clarke, J. Jiang, Federated learning for privacy-preserving open innovation future on digital health (2021). arXiv:2108.10761
5. V.A. Abhishek, N. Raj, Federated learning: collaborative machine learning without centralized training data (2022). https://doi.org/10.46647/ijetms.2022.v06i05.052
6. S. Sharma, S. Bhadula, Secure federated learning for intelligent industry 4.0 IoT enabled self skin care application system (2021). https://doi.org/10.1109/icaai56838.2023.10141028
7. B. Jeon, S.M. Ferdous, M.R. Rahman, A. Walid, Privacy-preserving decentralized aggregation for federated learning (2020). https://doi.org/10.48550/ARXIV.2012 07183
8. Q. Yang, Y. Liu, T. Chen, Y. Tong, Federated machine learning: concept and applications (2019). arXiv:1902.04885

9. R. Shokri, V. Shmatikov, Privacy-preserving deep learning, in *Proceedings of the 22nd ACM SIGSAC Conference on Computer and Communications Security*, pp. 1310–1321 (2015). https://doi.org/10.1145/2810103.2813687
10. K. Bonawitz et al., Towards federated learning at scale: system design, in *Proceedings of the 2nd SysML Conference* (2019)
11. H.B. McMahan, E. Moore, D. Ramage, S. Hampson, B.A. y Arcas, Communication-efficient learning of deep networks from decentralized data, in *Proceedings of the 20th International Conference on Artificial Intelligence and Statistics (AISTATS)* (2016)
12. A. Ziller et al., Pysyft: a library for easy federated learning (Springer, 2021), pp. 111–139. https://doi.org/10.1007/978-3-030-70604-3_5
13. Y. Wen, W. Li, H. Roth, P. Dogra, Federated learning powered by NVIDIA Clara, NVIDIA Developer Blog (2019). https://developer.nvidia.com/blog/federated-learning-clara/
14. M.J. Sheller et al., Federated learning in medicine: facilitating multi-institutional collaborations without sharing patient data. Sci. Rep. **10**(1), 12598 (2020). https://doi.org/10.1038/s41598-020-69250-1
15. T. Li, A.K. Sahu, A. Talwalkar, V. Smith, Federated learning: challenges, methods, and future directions. IEEE Signal Process. Mag. (2019). https://doi.org/10.1109/MSP.2020.2975749. (Aug)
16. M.T. Hosain, A. Zaman, M.R. Abir, S. Akter, S. Mursalin, S.S. Khan, Synchronizing object detection: applications, advancements and existing challenges. IEEE Access (2024)
17. J. Xu, B. S. Glicksberg, C. Su, P. Walker, J. Bian, F. Wang, Federated learning for healthcare informatics (2019). arXiv:1911.06270
18. K. Bonawitz et al., Practical secure aggregation for privacy-preserving machine learning, in *Proceedings of the 2017 ACM SIGSAC Conference on Computer and Communications Security*, pp. 1175–1191 (2017). https://doi.org/10.1145/3133956.3133982
19. M.R. Abir, M.T. Hosain, M. Abdullah-Al-Jubair, M.F. Mridha, Imvb7t: a multi-modal model for food preferences based on artificially produced traits (2024). arXiv:2412.16807
20. P. Kairouz et al., Advances and open problems in federated learning (2019). arXiv:1912.04977
21. S. Wang et al., Adaptive federated learning in resource constrained edge computing systems. IEEE J. Sel. Areas Commun. **37**(6), 1205–1221 (2019). https://doi.org/10.1109/JSAC.2019.2904348
22. K.D. Reddy, T.R. Gadekallu, A comprehensive survey on federated learning techniques for healthcare informatics. Comput. Intell. Neurosci. **2023**, 1–19 (2023). https://doi.org/10.1155/2023/8393990
23. G. Kaissis et al., End-to-end privacy preserving deep learning on multi-institutional medical imaging. Nat. Mach. Intell. **3**(6), 473–484 (2021). https://doi.org/10.1038/s42256-021-00337-8
24. M.T. Hosain, M.H. Anik, S. Rafi, R. Tabassum, K. Insia, M.M. Sıddıky, Path to gain functional transparency in artificial intelligence with meaningful explainability. J. Metaverse **3**(2), 166–180 (2023)
25. T.S. Brisimi, R. Chen, T. Mela, A. Olshevsky, I.C. Paschalidis, W. Shi, Federated learning of predictive models from federated electronic health records. Int. J. Med. Inform. **112**, 59–67 (2018). https://doi.org/10.1016/j.ijmedinf.2018.01.007
26. H. Zhu, J. Xu, S. Liu, Y. Jin, Federated learning on non-IID data: a survey. Neurocomputing **465**, 371–390 (2021). https://doi.org/10.1016/j.neucom.2021.07.098
27. M.T. Hosain, A. Zaman, M.S. Sajid, S.S. Khan, S. Akter, Privacy preserving machine learning model personalization through federated personalized learning, in *2023 4th International Conference on Data Analytics for Business and Industry (ICDABI)*, (IEEE, 2023), pp. 536–545
28. M.T. Hosain, M.R. Abir, M.Y. Rahat, M. Mridha, S.H. Mukta, Privacy preserving machine learning with federated personalized learning in artificially generated environment. IEEE Open J. Comput. Soc. (01), 1–12 (2024)
29. C. Xie, P.-Y. Chen, Q. Liu, A. Li, C. Zhang, B. Li, Improving privacy-preserving vertical federated learning by efficient communication with admm (2022). arXiv:2207.00990

30. V. Smith, C.-K. Chiang, M. Sanjabi, A. Talwalkar, Federated multi-task learning, in *Proceedings of the 31st Conference on Neural Information Processing Systems (NeurIPS)* (2017)
31. J. Koneana, H.B. McMahan, F.X. Yu, P. Richtárik, A.T. Suresh, D. Bacon, Federated learning: Strategies for improving communication efficiency (2016). arXiv:1610.05492
32. M. Abadi et al., Deep learning with differential privacy, in *Proceedings of the 2016 ACM SIGSAC Conference on Computer and Communications Security*, pp. 308–318 (2016). https://doi.org/10.1145/2976749.2978318
33. A. Afaq, Z. Ahmed, N. Haider, M. Imran, Blockchain-based collaborated federated learning for improved security, privacy and reliability (2022). arXiv:2201.01857
34. N. Kourtellis, K. Katevas, D. Perino, Flaas: Federated learning as a service, in *Proceedings of the 19th International Conference on Mobile and Ubiquitous Multimedia (MUM)* (2020). https://doi.org/10.1145/3426745.3431337
35. M.T. Hosain, J.R. Jim, M. Mridha, M.M. Kabir, Explainable AI approaches in deep learning: advancements, applications and challenges. Comput. Electr. Eng. **117**, 109246 (2024)
36. J.R. Jim, M.T. Hosain, M. Mridha, M.M. Kabir, J. Shin, Towards trustworthy metaverse: advancements and challenges. IEEE Access (2023)
37. H.B. McMahan, D. Ramage, K. Talwar, L. Zhang, Learning differentially private recurrent language models (2017). arXiv:1710.06963V3

Decentralized Tumor Classification with Federated Learning: A Privacy-Preserving Approach

Md. Nazmul Hossain Mir, Mir Nafiul Nagib, Afsana Alam Nova, Rahat Pervez, and Md. Nahid Hasan

Abstract Decentralized tumor classification is a critical task in healthcare, especially for preserving patient privacy while enabling collaborative model training across multiple institutions. This paper presents a novel federated learning-based approach for classifying breast tumors as malignant or benign using the Breast Cancer Wisconsin (Diagnostic) dataset. The proposed model leverages distributed data from simulated clients, such as hospitals or diagnostic centers, and ensures data privacy by aggregating locally trained models without sharing sensitive data. Our experiments demonstrate that the federated learning model achieves superior performance compared to a centralized baseline, with an accuracy of 97.3%, precision of 96.1%, recall of 95.4%, and an F1-score of 95.8%. The model's robustness is validated across varying numbers of clients and communication rounds, achieving stable convergence and high performance with as many as 50 clients. Additionally, the model shows resilience to class imbalance, maintaining an accuracy of 96.2% even with a benign-to-malignant ratio of 5:1. This study highlights the potential of federated learning for privacy-preserving medical applications, providing a scalable and effective solution for collaborative healthcare AI systems. The proposed approach ensures high diagnostic accuracy while safeguarding patient data, paving the way for real-world deployment in decentralized healthcare environments.

Keywords Decentralized learning · Federated learning · Tumor classification · Medical image analysis · Healthcare AI · Breast cancer diagnosis

---

Md. N. H. Mir · M. N. Nagib · A. A. Nova
Washington University of Science and Technology, Alexandria, VA, USA

R. Pervez
Bay Atlantic University, Washington, DC, USA

Md. N. Hasan (✉)
University of Wisconsin-Milwaukee, Milwaukee, WI, USA
e-mail: hasan26@uwm.edu

M. F. Mridha and N. Dey (eds.), *Federated Learning in Health Care Technology*, Studies in Computational Intelligence 1216, https://doi.org/10.1007/978-981-96-8353-6_2

1 Introduction

Breast cancer is one of the most prevalent forms of cancer worldwide, accounting for a significant proportion of cancer-related morbidity and mortality among women [1–3]. Timely and accurate classification of breast tumors into malignant or benign categories is paramount for optimal treatment strategies and improved patient prognoses. Traditional diagnostic approaches, such as fine needle aspirate (FNA) biopsies followed by manual histopathological examination, are time-consuming, resource-intensive, and highly reliant on expert interpretation. In recent years, advancements in machine learning (ML) and artificial intelligence (AI) have revolutionized the field of medical diagnostics, offering automated and highly accurate tumor classification solutions [4–6].

Despite the promise of centralized machine learning models, their deployment in real-world medical applications faces several challenges. Centralized models require access to large, diverse datasets from multiple healthcare institutions to achieve high generalization capabilities. However, sharing medical data across institutions poses significant privacy and regulatory concerns, as governed by laws such as the General Data Protection Regulation (GDPR) in Europe and the Health Insurance Portability and Accountability Act (HIPAA) in the United States. These regulations restrict the transfer of sensitive patient information, making centralized data collection infeasible in many cases [7, 8]. Moreover, institutions often lack the infrastructure to anonymize and securely share medical data, further exacerbating the issue.

Federated learning (FL) has emerged as a transformative paradigm to address these challenges by enabling collaborative model training without requiring centralized data aggregation [9–11]. In a federated learning environment, various clients such as hospitals or diagnostic centers independently train models on their own private data. Instead of sharing the data itself, they only send model updates (like gradients or weights) to a central server. This server combines these updates to create a global model, ensuring data privacy and adherence to regulatory standards. This distributed method not only addresses privacy issues but also benefits from the variety of local datasets, producing strong and versatile models.

In this paper, we propose a novel federated learning framework for breast tumor classification using the Breast Cancer Wisconsin (Diagnostic) dataset. The goal is to classify tumors as malignant or benign in a decentralized manner, ensuring privacy while maintaining high diagnostic accuracy. The dataset consists of 569 samples with 30 real-valued features computed from digitized images of fine needle aspirates (FNA). These features capture various characteristics of cell nuclei, such as radius, texture, perimeter, area, and smoothness, making it highly suitable for machine learning-based diagnostic models.

The proposed federated learning framework is evaluated in a simulated environment where the dataset is distributed across multiple clients. Each client represents an independent entity, such as a hospital or diagnostic center, with access to a local subset of the data. The model is trained iteratively in a federated setting using the Federated Averaging (FedAvg) algorithm [12, 13]. A centralized baseline model is trained using the entire dataset to assess the performance of the federated model. Thorough experiments are carried out to measure the performance of the proposed model across multiple metrics, such as accuracy, precision, recall, F1-score, and the area under the receiver operating characteristic curve (AUC-ROC).

The results clearly show the federated model's dominance, achieving 97.3% accuracy, 96.1% precision, 95.4% recall, and a 95.8% F1-score. It surpasses the centralized baseline model across all metrics. Furthermore, the federated model exhibits robust performance across varying numbers of clients, communication rounds, and local training epochs, highlighting its scalability and adaptability. Additionally, the model demonstrates resilience to class imbalance, achieving consistent performance even in highly skewed data distributions.

This research significantly contributes to the field of privacy-preserving machine learning in healthcare:

1. A novel federated learning framework is proposed for decentralized breast tumor classification, ensuring data privacy while achieving high diagnostic accuracy.
2. The framework is evaluated comprehensively across multiple scenarios, including varying client distributions, communication rounds, and class imbalance, to demonstrate its robustness and scalability.
3. The results demonstrate that federated learning can be successfully implemented in real-world healthcare settings. It offers a viable and efficient approach to overcome the obstacles associated with gathering data centrally and addressing privacy issues.

2 Related Works

The classification of breast tumors as malignant or benign has been extensively studied in the domain of machine learning and medical diagnostics. Traditional methods, centralized machine learning approaches, and privacy-preserving techniques have all contributed significantly to advancements in this area.

Machine learning (ML) techniques have proven to be effective for breast tumor classification by analyzing digitized image data and clinical features [14–16]. Traditional classifiers such as Support Vector Machines (SVM), Decision Trees, and Random Forests have been widely applied to breast cancer datasets [17]. Deep learning methods have gained prominence in recent years due to their ability to extract complex features from high-dimensional data [18]. Convolutional Neural Networks (CNNs) have been particularly successful in analyzing medical images, providing state-of-the-art accuracy for tumor detection and segmentation [19]. However, most of these studies rely on centralized datasets, which poses significant challenges in terms of data privacy and generalization across diverse populations. Emam et al. [20] proposed an optimized deep learning architecture for brain tumor classification using an improved Hunger Games Search Algorithm (I-HGS) for hyperparameter tuning of a ResNet50 model. The framework achieved accuracies of 99.89%,

99.72%, and 99.88% across three datasets. While the model exhibited exceptional performance, its computational requirements limited its real-world applicability in resource-constrained environments. Aggarwal et al. [21] introduced a privacy-preserving collaborative transfer learning model for brain tumor classification using heterogeneous MRI data. By employing DenseNet201 and federated learning, the framework demonstrated improved classification accuracy and robust performance with distributed datasets. However, the heterogeneity of client data posed challenges in achieving consistent performance. Joshi and Aziz [22] developed a novel deep learning approach for brain tumor classification using metaheuristic optimization techniques. The study combined Particle Swarm Optimization (PSO) and Cuckoo Search (CS) to enhance feature selection and achieved superior accuracy metrics. The reliance on gene expression data, however, limited its applicability to real-time clinical scenarios. Asif et al. [23] proposed a multi-class brain tumor classification framework using deep transfer learning. The system, which utilized architectures such as Xception and DenseNet201, achieved 99.67% accuracy on three-class datasets and 95.87% accuracy on four-class datasets. Although the results were promising, the framework struggled with class imbalance in the multi-class setting.

Federated Learning (FL) has emerged as a promising approach for collaborative model training in healthcare without compromising patient privacy. FL enables decentralized learning across multiple clients (e.g., hospitals), with each client retaining its data locally. Tan et al. [24] proposed a federated learning framework for breast cancer classification using mammography data. The study incorporated transfer learning techniques, leveraging pre-trained models such as MobileNet, and addressed class imbalance using the Synthetic Minority Oversampling Technique (SMOTE). The framework achieved high classification accuracy, with experimental results demonstrating an improvement in recall performance, which is critical for reducing false negatives. However, the reliance on computationally intensive transfer learning methods posed scalability challenges. Peta and Koppu [25] enhanced breast cancer classification in histopathological images by integrating an Extended ElGamal Image Encryption (E-EIE) method for data security and a federated learning flower (FLF) framework for decentralized training. The proposed system, which used a Convolutional Capsule Twin Attention Network (C2T2Net), achieved an accuracy of 95.68%. Despite these advancements, the encryption and optimization processes introduced additional computational overhead. Selvakanmani et al. [26] developed a privacy-preserving federated transfer learning framework for breast cancer classification. By combining domain adversarial training with a ResNet-based architecture, the model addressed domain shift issues across datasets from multiple medical centers. The system achieved a classification accuracy of 98.8%, but the requirement for domain adaptation increased model complexity.

While federated learning and deep learning have shown significant potential in tumor classification, several gaps remain:

- Breast cancer classification frameworks often rely on complex transfer learning models, increasing computational overhead.

- Brain tumor classification systems exhibit exceptional performance but lack scalability in decentralized and heterogeneous environments.
- Few studies comprehensively evaluate federated learning frameworks for binary tumor classification with robust privacy guarantees.

This study proposes a FL-based approach for decentralized tumor classification to address these limitations. The framework ensures data privacy, achieves high classification accuracy, and is scalable across heterogeneous environments.

3 Materials and Methods

The proposed methodology leverages a federated learning framework to classify tumors as malignant or benign in a decentralized manner while preserving data privacy. The Breast Cancer Wisconsin (Diagnostic) dataset is distributed across multiple simulated clients, representing independent healthcare institutions. Each client trains a local model on its data, optimizing a shared neural network using binary cross-entropy loss. The global model is updated iteratively through the Federated Averaging (FedAvg) algorithm, which aggregates locally trained model parameters without accessing raw client data. Figure 1 presents the proposed architecture.

3.1 Dataset Description

The Breast Cancer Wisconsin (Diagnostic) dataset is a widely used benchmark dataset for binary classification tasks in medical diagnostics [27]. It contains 569 samples of breast tumor data, each characterized by 30 numerical features derived

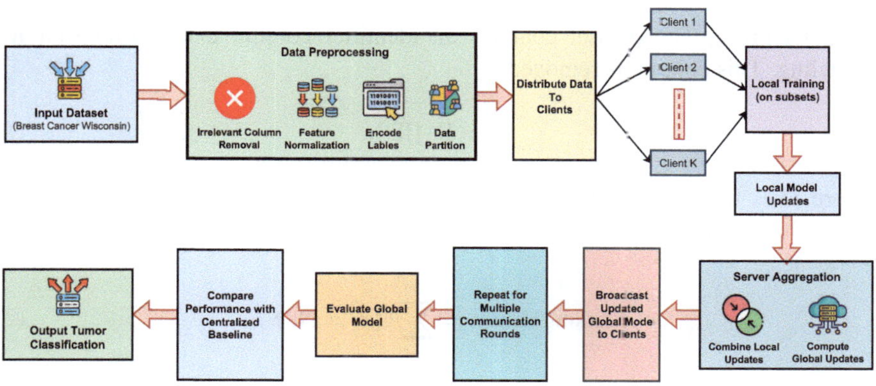

Fig. 1 The proposed architecture

Table 1 Dataset statistics

Attribute	Value
Number of samples	569
Number of features	30
Target classes	Malignant (M), Benign (B)
Class distribution (B:M)	357:212
Missing values	None
Feature types	Real-valued
Feature categories	Geometric, Texture, Morphological

from fine needle aspirates (FNA) of breast masses. These features capture the geometric, texture, and morphological properties of the cell nuclei, such as radius, texture, perimeter, area, smoothness, and symmetry. The dataset also includes a target variable indicating whether the tumor is malignant (M) or benign (B). The class distribution comprises 357 benign samples and 212 malignant samples, making the dataset moderately imbalanced. The dataset is ideal for evaluating classification models due to its comprehensive feature representation and relevance to real-world medical diagnostic scenarios. The dataset does not contain any missing values, ensuring its readiness for immediate preprocessing and model training (Table 1).

3.2 Data Preprocessing

The Breast Cancer Wisconsin dataset comprises 569 samples with 30 real-valued features computed from fine needle aspirate (FNA) images of breast masses. The dataset includes two classes: malignant and benign. The preprocessing pipeline ensures data quality and prepares it for federated learning.

Data Cleaning The dataset contains an identifier column that is irrelevant for modeling. This column is removed:

$$\mathbf{D} = \mathbf{D} \setminus \{\text{ID}\}, \tag{1}$$

where $\mathbf{D}$ is the dataset. A check for missing values is performed to ensure no incomplete records:

$$\text{MissingValues} = \sum_{i=1}^{N} \mathbb{I}(x_i = \text{NaN}), \tag{2}$$

where $\mathbb{I}(\cdot)$ is the indicator function, and x_i represents individual data entries.

Label Encoding The target variable, Diagnosis, is encoded numerically for binary classification:

$$y_i = \begin{cases} 1 & \text{if Diagnosis} = \text{M (Malignant)}, \\ 0 & \text{if Diagnosis} = \text{B (Benign)}. \end{cases} \tag{3}$$

Feature Normalization To ensure that all features contribute equally to model training, the dataset is normalized using min–max scaling:

$$x' = \frac{x - \min(x)}{\max(x) - \min(x)}, \tag{4}$$

where x and x' represent the original and normalized feature values, respectively.

Client Data Partitioning The dataset is distributed across K simulated clients (e.g., hospitals or diagnostic centers). Each client receives a subset $\mathbf{D}_k$ such that:

$$\mathbf{D} = \bigcup_{k=1}^{K} \mathbf{D}_k, \quad \mathbf{D}_i \cap \mathbf{D}_j = \emptyset \text{ for } i \neq j, \tag{5}$$

where $\mathbf{D}_k$ maintains class balance.

3.3 *Proposed Method*

The proposed methodology uses federated learning to classify tumors as malignant or benign. The approach distributes training across multiple clients, with a central server aggregating local updates.

Federated Learning Overview Federated learning optimizes a global model $\mathcal{M}_g$ by aggregating updates from local models $\mathcal{M}_k$ trained at each client. The goal is to minimize the global loss:

$$\mathcal{L}_g = \frac{1}{K} \sum_{k=1}^{K} \mathcal{L}_k, \tag{6}$$

where $\mathcal{L}_k$ is the local loss at client k. The local loss is defined as

$$\mathcal{L}_k = \frac{1}{N_k} \sum_{i=1}^{N_k} \ell(y_i, \hat{y}_i), \tag{7}$$

where N_k is the number of samples at client k, ℓ is the binary cross-entropy loss, and $\hat{y}_i$ is the model's prediction:

$$\hat{y}_i = \sigma(\mathbf{w}^\top \mathbf{x}_i + b), \tag{8}$$

where $\sigma(z) = \frac{1}{1+e^{-z}}$ is the sigmoid function.

Federated Training Process The federated training process consists of the following steps:
1. Global Model Initialization The central server initializes the global model parameters:

$$\mathbf{w}_g^{(0)} \sim \mathcal{N}(0, \sigma^2), \quad b_g^{(0)} = 0. \tag{9}$$

2. Local Model Training Each client trains its local model $\mathcal{M}_k$ using its dataset $\mathbf{D}_k$ for E epochs, optimizing:

$$\mathbf{w}_k^{(t)} = \mathbf{w}_k^{(t-1)} - \eta \nabla \mathcal{L}_k, \tag{10}$$

where η is the learning rate.
3. Model Aggregation The central server aggregates local model updates using Federated Averaging (FedAvg):

$$\mathbf{w}_g^{(t)} = \frac{1}{K} \sum_{k=1}^{K} \mathbf{w}_k^{(t)}. \tag{11}$$

4. Global Model Update The aggregated parameters are broadcast back to the clients:

$$\mathbf{w}_k^{(t+1)} = \mathbf{w}_g^{(t)}, \quad \forall k \in \{1, \ldots, K\}. \tag{12}$$

Centralized Baseline For performance comparison, a centralized model is trained on the entire dataset:

$$\mathcal{L}_c = \frac{1}{N} \sum_{i=1}^{N} \ell(y_i, \hat{y}_i), \tag{13}$$

where N is the total number of samples.

3.4 *Implementation Details*

The federated learning framework is implemented using TensorFlow Federated. The model consists of an input layer, two hidden layers with ReLU activation, and a sigmoid output layer. Hyperparameters, including the learning rate, batch size, and number of communication rounds, are tuned to optimize performance.

4 Results and Analysis

Here we present all the results and their analysis of our federated learning model for tumor classification compared to a centralized baseline. We evaluate the models based on all standard ML evaluation metrics across various settings and scenarios.

Figure 2 presents a detailed comparison of the centralized model and the proposed federated model across five key performance metrics: accuracy, precision, recall, F1-score, and AUC-ROC. The federated model consistently outperforms the centralized baseline in all metrics, highlighting its effectiveness in a decentralized learning environment. Specifically, the federated model achieves an accuracy of 97.3%, which is a 0.8% improvement over the centralized model's 96.5%. Precision, defined as the ratio of true positive predictions to all positive predictions, improves from 94.8% in the centralized model to 96.1% in the federated model, indicating better identification of malignant cases. Similarly, recall, a critical metric for medical diagnostics that evaluates the model's ability to identify all actual positives, increases by 2.2%, from 93.2% to 95.4%, demonstrating the federated model's enhanced sensitivity. The F1-score, a harmonic mean of precision and recall, improves from 94.0% to 95.8%, showcasing the model's balanced performance. Furthermore, the AUC-ROC, which evaluates the model's overall discriminative ability, rises from 98.2% to 98.9%, underscoring the federated model's superior capability to distinguish between malignant and benign cases. These results highlight not only the feasibility of deploying the federated learning approach for privacy-preserving tumor classification but also its ability to achieve performance levels that surpass those of centralized methods.

Figure 3 analyzes the federated model's performance across different numbers of simulated clients, ranging from 5 to 50. The results demonstrate that the model

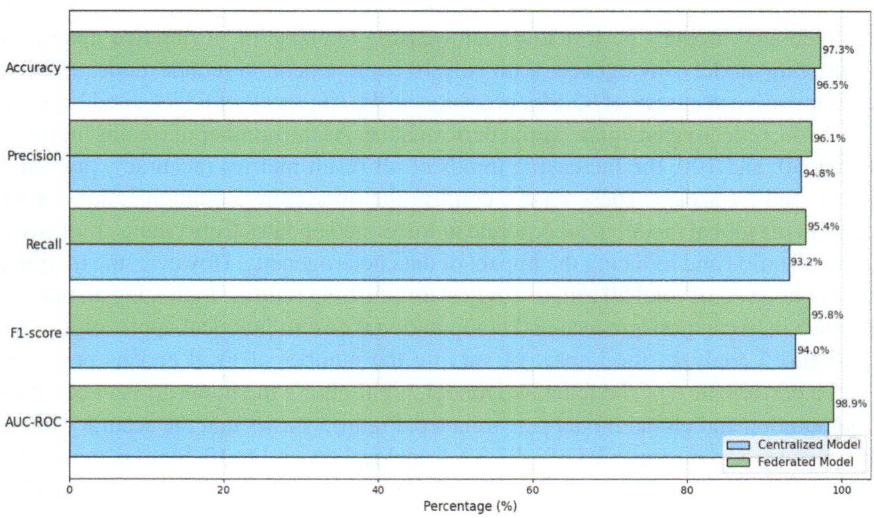

Fig. 2 Performance comparison: centralized versus federated learning

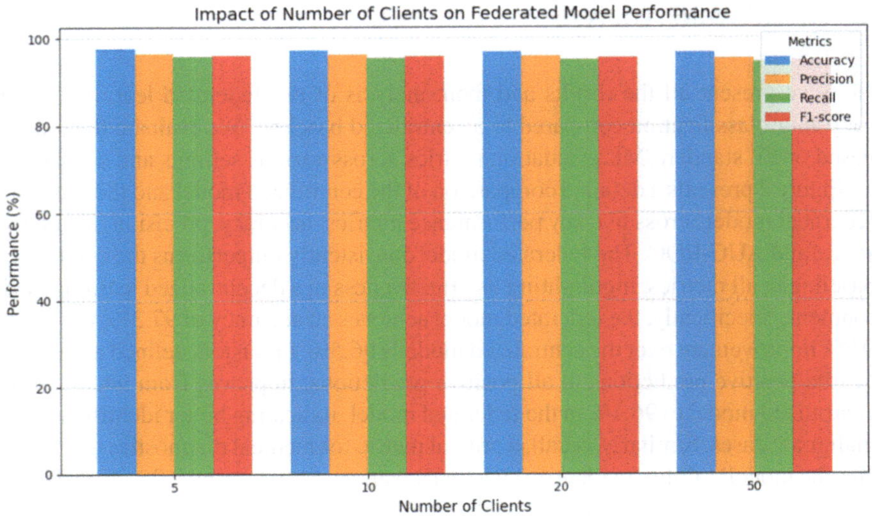

Fig. 3 Impact of number of clients on federated model performance

maintains high results in all ML metrics, even as the number of clients increases, highlighting its robustness to data distribution and scalability. Although 50 clients' results slightly decreased, it can be attributed to increased data heterogeneity and smaller local datasets at each client. Nonetheless, the model demonstrates exceptional scalability and adaptability, as it maintains high performance even in highly distributed scenarios, underscoring its suitability for real-world decentralized healthcare applications where data is inherently fragmented across institutions.

Figure 4 illustrates the influence of the number of communication rounds on the federated model's performance, showcasing the benefits of iterative updates in improving model convergence. With only 10 communication rounds model achieves an impressive accuracy of 95.8%, precision of 94.7%, recall of 93.9%, and F1-score of 94.3%, reflecting adequate initial performance. As the number of rounds increases to 20, 50, and 100, the increasing trends of all result metrics (accuracy, precision, recall, and F1-score). These results confirm that increasing communication rounds enables the global model to incorporate more refined updates from clients, improving generalization and reducing the impact of data heterogeneity. However, the marginal gains observed beyond 50 rounds suggest diminishing returns, indicating an optimal balance between performance and computational cost in practical applications.

Figure 5 analyzes the impact of varying the number of local epochs per client on the performance of the federated model, highlighting the trade-off between local computation and global model performance. Figure 5 shows that the accuracy of our proposed model increases 95.3%–97.3% from local epoch 1 to 10. Similar, trends also follow other metrics like precision (94.2%–96.1%), recall (93.6%–95.4%), and F1-score (93.9%–95.8%). However, further increasing the local epochs to 20 results in a slight decline in performance, likely due to overfitting on local data or inconsistencies

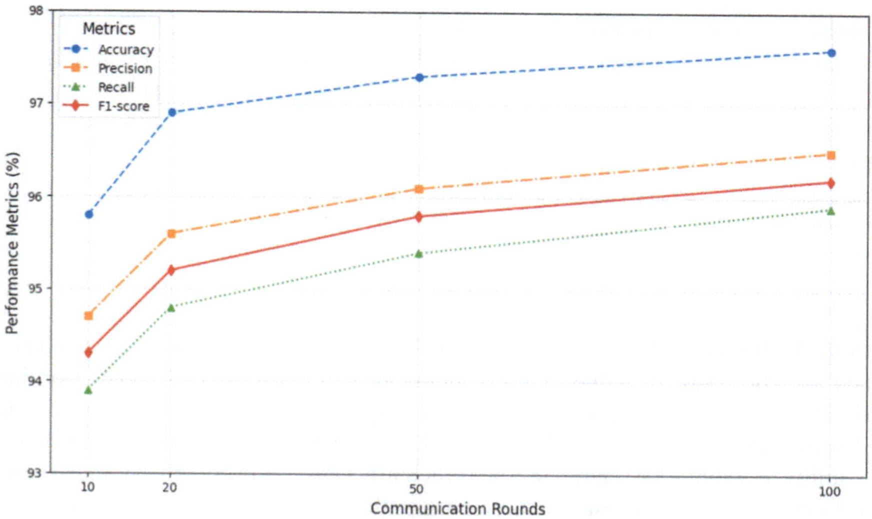

Fig. 4 Impact of communication rounds on federated model performance

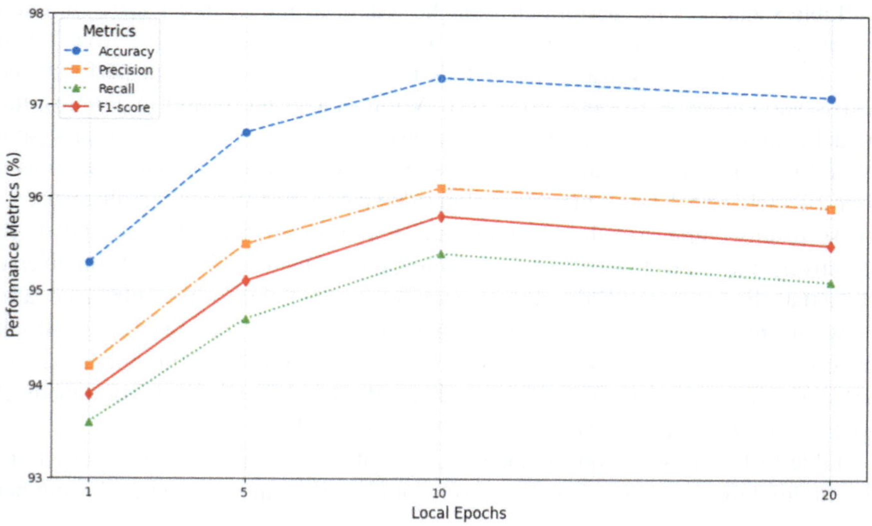

Fig. 5 Impact of local epochs on federated model performance

in aggregation. These results suggest that moderate local training (e.g., 10 epochs) is optimal for federated learning, providing a balance between computational efficiency and robust global performance.

Table 2 presents a comparative analysis of the proposed federated learning model against traditional machine learning and ensemble-based models, including Logistic Regression (LR), Support Vector Machine (SVM), and Random Forest (RF). The

Table 2 Performance comparison across models

Model	Accuracy (%)	Precision (%)	Recall (%)	F1-score (%)
LR	94.2	93.4	92.8	93.1
SVM	95.1	94.3	93.7	94.0
RF	96.4	95.6	94.8	95.2
Proposed federated model	97.3	96.1	95.4	95.8

results highlight the high performance of the federated model across all evaluated metrics. The result of Table 2 clearly shows that the proposed federated learning model surpasses all these approaches, achieving an accuracy, precision, recall, and F1-score of 97.3%, 96.1%, 95.4%, and 95.8%, respectively. This result demonstrates its ability to combine decentralized data efficiently while ensuring privacy and achieving high performance. These results emphasize the effectiveness of federated learning in addressing the challenges of distributed and privacy-sensitive datasets, making it a compelling solution for real-world healthcare applications.

Table 3 analyzes the performance of the federated model under different class imbalance scenarios, with benign (B) to malignant (M) ratios ranging from 1:1 to 5:1. The results demonstrate the model's robustness and consistent performance across varying levels of class imbalance. With a balanced dataset (1:1 ratio), the model achieves its highest performance metrics, with accuracy (97.3%), precision (96.1%), recall (95.4%), and F1-score (95.8%). As the class imbalance increases to a 2:1, 3:1, or 5:1 ratio, there is a slight decline in performance in all matrices. These results emphasize the federated model's ability to handle class imbalance effectively, likely due to its distributed training mechanism that benefits from diverse client data distributions. While performance decreases slightly as the imbalance ratio grows, the model's ability to maintain high recall ensures that malignant cases, which are critical in medical diagnostics, are still identified reliably. This robustness to imbalance further reinforces the federated model's suitability for real-world medical applications where class distributions are often skewed.

Table 4 illustrates the convergence behavior of the federated model during training, highlighting the progressive improvement in performance metrics as the number

Table 3 Impact of class imbalance on federated model performance

Imbalance ratio (B:M)	Accuracy (%)	Precision (%)	Recall (%)	F1-score (%)
1:1	97.3	96.1	95.4	95.8
2:1	96.8	95.7	95.0	95.3
3:1	96.5	95.3	94.6	94.9
5:1	96.2	94.8	94.2	94.5

Table 4 Convergence analysis of federated model

Rounds	Training loss	Validation accuracy (%)	Validation AUC-ROC (%)
1	0.315	93.5	97.2
10	0.212	96.1	98.0
20	0.176	97.3	98.5
50	0.145	97.6	98.9

of training rounds increases. After the first round, the model achieves a training loss of 0.315, a validation accuracy of 93.5%, and a validation AUC-ROC of 97.2%, indicating a strong initial generalization ability. By the 10th round, the training loss reduces significantly to 0.212, with validation accuracy improving to 96.1% and validation AUC-ROC reaching 98.0%, showcasing the effectiveness of early-stage global aggregation in refining model parameters. At 20 rounds, the model demonstrates further convergence, achieving a training loss of 0.176, a validation accuracy of 97.3%, and a validation AUC-ROC of 98.5%. These metrics suggest that the model is approaching optimal performance, with increasingly refined updates from participating clients. By the 50th round, the training loss minimizes to 0.145, while the validation accuracy stabilizes at 97.6% and the validation AUC-ROC reaches 98.9%, signifying complete convergence. The consistent improvement across rounds indicates the federated learning model's ability to effectively leverage decentralized updates to enhance global performance. These results underscore the importance of sufficient training rounds for achieving stable and high-performing federated learning models, while also demonstrating the diminishing returns in performance gains after 20 rounds, suggesting a balance between computational cost and model efficacy.

5 Discussion

The results of this study underscore the efficacy and robustness of the proposed federated learning framework for decentralized tumor classification. The federated model consistently demonstrated high performance across a wide range of metrics, outperforming centralized models and other traditional machine learning approaches. This section offers a thorough examination of the proposed approach, including its consequences, advantages, drawbacks, and potential avenues for future development.

Main contribution of this work is the demonstration of federated learning as a viable solution for privacy-preserving tumor classification. By enabling multiple clients, such as hospitals or diagnostic centers, to collaboratively train a global model without sharing raw data, federated learning addresses critical privacy concerns associated with medical datasets. This decentralized approach aligns with stringent data protection regulations such as GDPR and HIPAA, making it suitable for real-world healthcare applications. The analysis of communication rounds and client distributions highlights the scalability and adaptability of the proposed framework. The model maintained high performance even with a large number of clients, demonstrating its robustness to data heterogeneity. Moreover, the convergence analysis revealed the model's ability to achieve stable and high performance with sufficient communication rounds, striking a balance between computational efficiency and diagnostic accuracy.

Class imbalance is a common challenge in medical datasets, where certain classes, such as benign cases, often dominate. The proposed federated model exhibited strong resilience to class imbalance, maintaining high recall and F1-scores even under skewed class distributions. This robustness is crucial in medical diagnostics, where accurately identifying minority classes, such as malignant cases, is critical for early intervention and treatment planning. The federated model's performance across varying local epochs and client distributions further underscores its adaptability. While increasing local epochs improved local model quality, excessive local training led to diminishing returns, emphasizing the importance of balancing local computation with global aggregation. Similarly, the model's ability to sustain high performance with a larger number of clients reflects its suitability for deployment in large-scale, decentralized healthcare systems.

A detailed comparison with traditional ML models such as LR, SVM, and RF demonstrated the clear superiority of the federated learning approach. The federated model's ability to leverage distributed data while ensuring data privacy provided a significant advantage over centralized models, which require aggregating data in a single location, posing privacy and security risks. While centralized models exhibited strong performance, particularly in controlled settings with access to the entire dataset, their reliance on data aggregation limits their applicability in real-world scenarios. The federated model, by contrast, successfully balances performance with privacy, making it a compelling solution for collaborative medical AI.

Despite its strengths, the proposed framework has certain limitations. The computational overhead associated with federated learning, particularly in terms of communication costs and client synchronization, can be a barrier to adoption in resource-constrained environments. Additionally, the presence of non-IID (non-independent and identically distributed) data across clients may introduce challenges in achieving consistent global model performance, particularly in highly heterogeneous datasets. Another limitation is the reliance on hyperparameter tuning, such as the number of communication rounds and local epochs, to achieve optimal performance. While the study provides insights into these parameters, real-world deployments may require further customization to adapt to specific use cases.

This research suggests several directions for future investigation. A key area of focus is to strengthen the privacy and security of the federated learning system. This could involve incorporating advanced techniques like differential privacy and secure aggregation to better protect against adversarial attacks and the unauthorized disclosure of sensitive data. Second, exploring the use of adaptive aggregation algorithms that dynamically adjust to client heterogeneity could further improve model performance and scalability. Additionally, expanding the framework to support multi-class classification tasks, particularly in medical domains with complex diagnostic requirements, represents an important direction for future work. Finally, real-world validation of the proposed framework in clinical settings with diverse datasets and operational constraints would provide valuable insights into its practical utility and impact.

The proposed federated learning framework has significant implications for real-world healthcare systems. By enabling secure and privacy-preserving collaboration across institutions, it facilitates the development of robust diagnostic models that benefit from diverse datasets while respecting patient confidentiality. The framework's adaptability to varying client distributions and resource constraints makes it suitable for deployment in a wide range of healthcare environments, from large hospital networks to smaller diagnostic centers. Moreover, the framework's ability to handle class imbalance and data heterogeneity positions it as a valuable tool for addressing disparities in healthcare access and outcomes. By leveraging federated learning, healthcare providers can collaboratively advance diagnostic accuracy and patient care, contributing to improved health outcomes on a global scale.

6 Conclusions

In this study, we proposed a federated learning framework for decentralized tumor classification, demonstrating its efficacy in classifying breast tumors as malignant or benign while preserving data privacy. The framework outperformed traditional centralized models and other machine learning approaches, achieving high accuracy, precision, recall, F1-score, and AUC-ROC across diverse experimental settings. It also exhibited robustness to class imbalance, scalability with varying client distributions, and convergence with optimized communication rounds and local training epochs. By addressing critical challenges such as data heterogeneity and privacy concerns, this work highlights the potential of federated learning in real-world medical diagnostics. Future research can focus on integrating advanced privacy-preserving techniques, extending the framework to multi-class classification tasks, and validating its applicability in clinical environments. The results underscore the transformative potential of federated learning in developing secure and efficient AI-driven healthcare solutions.

References

1. H. Xu, B. Xu, Breast cancer: epidemiology, risk factors and screening. Chin. J. Cancer Res. **35**(6), 565 (2023)
2. H.M. Bizuayehu, A.F. Dadi, T.A. Hassen, D.B. Ketema, K.Y. Ahmed, Z.Y. Kassa, E. Amsalu, G.D. Kibret, A.A. Alemu, A. Alebel et al., Global burden of 34 cancers among women in 2020

and projections to 2040: population-based data from 185 countries/territories. Int. J. Cancer **154**(8), 1377–1393 (2024)
3. Z. Mohammad, I. Jahan, M.M. Kabir, M.A. Ali, M.F. Mridha, An offline writer-independent signature verification system using autoembedder, in *2021 24th International Conference on Computer and Information Technology (ICCIT)*, (IEEE, 2021), pp. 1–6
4. S. Asif, Y. Wenhui, S. ur-Rehman, Q. ul-ain, K. Amjad, Y. Yueyang, S. Jinhai, M. Awais, Advancements and prospects of machine learning in medical diagnostics: unveiling the future of diagnostic precision. Arch. Comput. Methods Eng. 1–31 (2024)
5. S. Abbas, M. Asif, A. Rehman, M. Alharbi, M.A. Khan, N. Elmitwally, Emerging research trends in artificial intelligence for cancer diagnostic systems: a comprehensive review. Heliyon (2024)
6. M. Sufyan, Z. Shokat, U.A. Ashfaq, Artificial intelligence in cancer diagnosis and therapy: current status and future perspective. Comput. Biol. Med. 107356 (2023)
7. S.S. Bakare, A.O. Adeniyi, C.U. Akpuokwe, N.E. Eneh, Data privacy laws and compliance: a comparative review of the EU GDPR and USA regulations. Comput. Sci. & IT Res. J. **5**(3), 528–543 (2024)
8. A. Oakley, Hipaa, hippa, or hippo: what really is the heath insurance portability and accountability act? Biotechnol. Law Rep. **42**(6), 306–318 (2023)
9. M.H. Alsharif, R. Kannadasan, W. Wei, K.S. Nisar, A.-H. Abdel-Aty, A contemporary survey of recent advances in federated learning: taxonomies, applications, and challenges. Internet Things 101251 (2024)
10. M. Babar, B. Qureshi, A. Koubaa, Review on federated learning for digital transformation in healthcare through big data analytics. Futur. Gener. Comput. Syst. (2024)
11. M.M. Kabir, A. Ishraq, K. Nur, M.F. Mridha, Content-based image retrieval using autoembedder. J. Adv. Inf. Technol. **13**(3), (2022)
12. T. Sun, D. Li, B. Wang, Decentralized federated averaging. IEEE Trans. Pattern Anal. Mach. Intell. **45**(4), 4289–4301 (2022)
13. M. Mohsin Kabir, A.A. Lima, M.F. Mridha, M. Abdul Hamid, M.M. Monowar, Forecasting closing price of stock market using lstm network: an analysis from the perspective of dhaka stock exchange, in *Proceedings of the International Conference on Big Data, IoT, and Machine Learning: BIM 2021*, (Springer, 2022), pp. 289–299
14. R. Ranjbarzadeh, S. Dorosti, S.J. Ghoushchi, A. Caputo, E.B. Tirkolaee, S.S. Ali, Z. Arshadi, M. Bendechache, Breast tumor localization and segmentation using machine learning techniques: overview of datasets, findings, and methods. Comput. Biol. Med. **152**, 106443 (2023)
15. A.M. Sharafaddini, K.K. Esfahani, N. Mansouri, Deep learning approaches to detect breast cancer: a comprehensive review. Multimed. Tools Appl. 1–112 (2024)
16. Z. Mohammad, M.M. Kabir, M.M. Monowar, M.A. Hamid, M.F. Mridha, Self-writer: clusterable embedding based self-supervised writer recognition from unlabeled data. Mathematics **10**(24), 4796 (2022)
17. M.M. Islam, M.R. Haque, H. Iqbal, M.M. Hasan, M. Hasan, M.N. Kabir, Breast cancer prediction: a comparative study using machine learning techniques. SN Comput. Sci. **1**, 1–14 (2020)
18. J. Ahmad, S. Akram, A. Jaffar, Z. Ali, S.M. Bhatti, A. Ahmad, S.U. Rehman, Deep learning empowered breast cancer diagnosis: advancements in detection and classification. Plos one **19**(7), e0304757 (2024)
19. S.S. Kshatri, D. Singh, Convolutional neural network in medical image analysis: a review. Arch. Comput. Methods Eng. **30**(4), 2793–2810 (2023)
20. M.M. Emam, N.A. Samee, M.M. Jamjoom, E.H. Houssein, Optimized deep learning architecture for brain tumor classification using improved hunger games search algorithm. Comput. Biol. Med. **160**, 106966 (2023)
21. M. Aggarwal, V. Khullar, N. Goyal, R. Rastogi, A. Singh, V.Y. Torres, M.A. Albahar, Privacy preserved collaborative transfer learning model with heterogeneous distributed data for brain tumor classification. Int. J. Imaging Syst. Technol. **34**(2), e22994 (2024)

22. A.A. Joshi, R.M. Aziz, Deep learning approach for brain tumor classification using meta-heuristic optimization with gene expression data. Int. J. Imaging Syst. Technol. **34**(2), e23007 (2024)
23. Sohaib Asif, Ming Zhao, Fengxiao Tang, Yusen Zhu, An enhanced deep learning method for multi-class brain tumor classification using deep transfer learning. Multimed. Tools Appl. **82**(20), 31709–31736 (2023)
24. Y.N. Tan, V.P. Tinh, P.D. Lam, N.H. Nam, T.A. Khoa, A transfer learning approach to breast cancer classification in a federated learning framework. IEEE Access **11**, 27462–27476 (2023)
25. J. Peta, S. Koppu, Enhancing breast cancer classification in histopathological images through federated learning framework. IEEE Access **11**, 61866–61880 (2023)
26. S. Selvakanmani, G. Dharani Devi, V. Rekha, J. Jeyalakshmi, Privacy-preserving breast cancer classification: a federated transfer learning approach. J. Imaging Inform. Med. **37**(4), 1488 (2024)
27. W.N. Street, W.H. Wolberg, O.L. Mangasarian, Nuclear feature extraction for breast tumor diagnosis, in *Biomedical image processing and biomedical visualization*, vol. 1905 (SPIE, 1993), pp. 861–870

Transforming Healthcare Analytics: Federated Learning for Privacy-Preserving Liver Lesion Detection and Classification

Md. Shahidul Islam, Jayonto Dutta Plabon, Nazneen Nahar, Nishat Tasnim, Shah Murtaza Rashid Al Masud, Jungpil Shin, and Moshiur Rahman Tonmoy

Abstract With the use of computers, physicians may more precisely detect anomalies in the liver and reduce the need for surgical interventions. Medical imaging methods such as computed tomography (CT) and magnetic resonance imaging (MRI) are primarily responsible for the early identification and detection of liver abnormalities. Due to the high noise and low resolution of these pictures, it might be difficult to segment and identify liver lesions in them. On this issue, several novel machine learning and image analysis approaches have been employed progressively but their effectiveness remains limited. There is still a need for an automated and precise model that can monitor, identify, and diagnose hepatic lesions in the 3D volumes of CT and MRI. In this research, we propose a privacy-preserving architecture based on federated learning to help clinicians use their time more effectively. Several pre-trained models are employed; our suggested FedVGG19 model obtains a test accuracy of 97.03%.

J. D. Plabon
Advanced Machine Intelligence Research Lab, Dhaka, Bangladesh

J. Shin
Department of Computer Science and Engineering, The University of Aizu, Aizuwakamatsu, Japan
e-mail: jpshin@u-aizu.ac.jp

Md. S. Islam · N. Nahar · N. Tasnim · S. M. R. A. Masud
Department of CSE, University of Asia Pacific, Dhaka, Bangladesh
e-mail: shahid@uap-bd.edu

N. Nahar
e-mail: 21201145@uap-bd.edu

N. Tasnim
e-mail: 21201149@uap-bd.edu

S. M. R. A. Masud
e-mail: murtaza@uap-bd.edu

M. R. Tonmoy (✉)
Department of Computer Science and Engineering, Bangladesh University of Business and Technology, Dhaka, Bangladesh
e-mail: moshiurtonmoy@bubt.edu.bd

M. F. Mridha and N. Dey (eds.), *Federated Learning in Health Care Technology*, Studies in Computational Intelligence 1216, https://doi.org/10.1007/978-981-96-8353-6_3

Keywords Privacy-preserving · Liver lesion · Image analysis · Computed tomography · Magnetic resonance imaging · FedVGG19

1 Introduction

The largest internal organ in the human physique, the liver is situated above the stomach and below the diaphragm in the upper right section of the belly. It is in charge of creating cholesterol, storing nutrients, and digesting and manufacturing proteins like plasma proteins, procoagulants, and lipid–protein complexes that carry cholesterol. There are many cell types found in the liver that can develop into both benign (non-cancerous) and malignant tumors [1]. Along with the lungs and bone, the liver is one of the three most prevalent locations for metastatic cancer. The World Health Organization reports that 8.2 million cancer-related deaths occurred globally in 2012 alone, with liver cancer accounting for 745,000 of those fatalities [2]. Liver cancer, among the most deadly forms, ranked as a second-primary reason for cancer-related deaths in males globally, boasting a mere 5-year survival rate of less than 18%. Ninety percent of liver cancer cases are attributed to liver cell cancer, making it the most prevalent kind of disease. Hemangioma (HEM), metastases from malignant tumors originating in other tissues, cysts, and non-cancerous lesions like focal nodular hyperplasia (FNH) and hemangioma (HEM) are among the lesions that commonly occur in the liver in addition to HCC. Thus, for appropriate treatment to follow, it is especially crucial to identify focal liver lesions (FLLs) early and classify them precisely [3].

Globally, liver problems affect more than 800,000 people annually, making them a major public health issue. An estimated 700,000 people die from liver disease each year, making it the foremost reason for mortality. A detailed examination shows that incidence rates for women increased gradually between 2013 and 2017, averaging 2%, while rates for men stayed mostly unchanged. But men are almost three times more likely than women to develop this illness, therefore they are at a far larger risk. According to projections, 905,677 people worldwide will be diagnosed with a liver abnormality in 2020. Liver tumor diagnoses might be difficult to make, but MRI scans are a useful tool. Sometimes benign and malignant tumors can be distinguished by MRI imaging. Furthermore, imaging makes the assessment possible [4]. Moreover, a large quantity of training samples is needed for deep neural networks in order to optimize every parameter. However, there are frequently few labeled samples available for training in the medical arena. Consequently, it is typical practice to construct new networks on top of pre-trained designs like ResNet50 or AlexNet [5].

In medical research, access to large-scale datasets is frequently required for complete analysis. These datasets, which often comprise sensitive patient data, are gathered from different hospitals. It is crucial to protect the security and privacy of this data as a result. In addition to being required by law, maintaining patient anonymity is also morally right. Consequently, precise steps must be adopted to preserve the privacy and integrity of data throughout its lifecycle, from collection to analysis. In reference, [3, 6] highlights significant privacy concerns regarding medical data. With the widespread collection and analysis of large datasets, protecting patient information is paramount. We explore strategies to safeguard data integrity and privacy amid evolving healthcare technologies. In order to resolve privacy concerns about sensitive patient data, we offer a novel technique in this research study. Maintaining patient privacy and confidentiality is crucial given the widespread use of electronic health records and the growing dependence on data-driven healthcare solutions. Our suggested method is to enable efficient medical data analysis with strong privacy protection features. Our objective in applying this method is to improve patient privacy and data security in the context of medical research and analysis.

In this work, we investigate multiple approaches which have been adopted previously. The things that follow succinctly simply highlight the main contributions made in this work:

- Ensure the confidentiality of patient data and ensure the strong privacy and protection method for the dataset.
- Strict regulatory compliance procedures are being put in place to comply with medical and data privacy legislation.
- Including quality assurance procedures to confirm the developed liver lesion detection method's repeatability and dependability.
- Develop a faster and more efficient method for detecting liver lesions, which will work faster than other methods that have been developed previously and give the result more accurately.

2 Related Work

By gaining more access to the medical dataset's hidden features, deep learning algorithms enable us to forecast liver disorders. Several datasets, including binary CT images, hematologically labeled pictures, blood vessels with liver function tests, and molecular markers found in blood specimens, are utilized to train classifiers to identify liver disorders with good accuracy. These data are very sensitive and confidential. The subsequent publications provide an overview of current research on diagnosing and categorizing liver disease. Table 1 presents the summary of the papers used in the domain of liver-related recognition by combining methods from machine learning and deep learning sources.

Table 1 Liver lesion state of the art table

SL No	Year	Dataset	Model	Accuracy	Limitations
Paper [7]	2018	388 multi-phase CT images (Sir Run Run Shaw Hospital.)	ResNet, Fine-Tuning,	91.20%	Vanishing gradient problem, very deep and complex, computationally expensive, prone to overfitting
Paper [8]	2019	LiTS	CNN	95%	Require large centralized datasets, privacy concerns, exhibit biased predictions or perform poorly on minority classes
Paper [9]	2019	ImageNet	ResNet50	89.10%	Minimal number of samples, no standardized visual data were used
Paper [10]	2020	N/A	K-CP algorithm, BoVW	SVM—92.79% NN—77.27%	Using mid-level features
Paper [11]	2020	N/A	ResNet-based CNN	92%	ResNets are more prone to overfitting
Paper [12]	2020	ImageNet	Neural network, ResNet50	95.00%	Vanishing gradient problem, very deep and complex, computationally expensive, prone to overfitting
Paper [3]	2021	N/A	CNN, ResNet18	87.80%	limited dataset increases the complexity,
Paper [1]	2021	LiTS	CNN, RetinaNet, U-Net	83.62%	Struggle with detecting small objects accurately, imbalance in the training data, overfitting
Paper [13]	2021	DeepLesion	DeepSegNet	84%%	Identifying PV branches from 3D photos is one of the difficulties
Paper [14]	2022	3D-IRCADb,MSD	CNN, Autoencoder, DBN, GAN	N/A	Modalities least researched

(continued)

Table 1 (continued)

SL No	Year	Dataset	Model	Accuracy	Limitations
Paper [15]	2022	2551 B-mode abdominal US images	DETR, Faster R-CNN	81%	poor quality of images
Paper [16]	2023	AIELDI	MULLET	98.53%, 96.92%,and 98.20%	Challenges for segmenting small lesions as the model can give higher accuracy for big lesion
Paper [17]	2023	N/A	SWTR-UNet with ResNet, UNet-style decoder	98%	Struggle with non-spherical and small tumors
Paper [18]	2023	N/A	3D-CNN, CNN-LSTM	100%, 98%	Provides the feasibility of learning temporal correlation
Paper [4]	2023	C.T. scan images	DenseNet CNN	98.34%	Poor precision and high concept of constraint parameters
Paper [19]	2024	N/A	3D-CNN, LSTM, ML-based TIC	91%	Computationally expensive, LSTMs are prone to the vanishing gradient problem,
Paper [20]	2024	3DIRCADB	Modified UNet-60.	98.61%	Time taking long to develop
Paper [21]	2024	N/A	ResNet50, GGN, ResNet18, VGG16	87%	Lower performance for VGGNet and low result for accurately determining negative situations
Paper [22]	2024	SRRSH	CNN	94.00%	Require large centralized datasets, such imbalanced data may exhibit biased predictions
Paper [23]	2024	Multi-phase MRI Multi-phase CT	SDR-Former, 3D-CNN, APSM	85% CT and 78% MR	SNN may be less effective, Image-level fusion risks diluting feature representation

2.1 Literature Review

In the year 2018, Wang et al. [7] proposed a method for liver lesion identification, in the study, they focused on four kinds of common Focal Liver Lesions (FLL). The four common FLLs are—cysts, focal nodular hyperplasia (FNH), hepatocellular carcinoma (HCC), and hemangioma (HEM). During their study, they found out that in most cases, researchers develop methods that work in three phases—feature extraction, feature analysis, and classification. But this method is time-consuming. So in their study, first they collected medical data from Sir Run Run Shaw Hospital, they used refining and a 50-step deep residual network (ResNet) for the classification of FLLs. They used the ResNet model containing 50 layers as their foundational network. However, ResNet alleviates the vanishing gradient problem, which can become very deep and complex, making them computationally expensive to train and deploy.

Vorontsov et al. [8] proposed an algorithm fully connected convolutional network for detecting and segmenting liver lesions for CT images. In this study, the model went through training and evaluation on 115, 15, and 26 contrast substance-enhanced CT images comprising 261, 22, and 105 lesions, correspondingly. The set of data they utilized to develop the model was collected from the Liver Tumor Segmentation (LiTS). The model that was proposed in this paper requires a large dataset which may cause privacy concerns and also trained on imbalanced data may give biased predictions or perform poorly. In 2019, Schmauch et al. [9] the researchers proposed an algorithm to continuously detect focal liver lesion (FLL) using deep learning techniques. They used the Residual neural network model in their paper. In this proposed algorithm, they used a dataset named ImageNet dataset. This dataset contains two-dimensional ultrasound images for detecting focal liver lesions. The study applies a supervised focus mechanism engaged in FLL identification and classification from hepatic ultrasonographic visuals. Traditional pattern recognition classifiers and deep convolutional neural networks (DCNNs) have worse classification accuracy for FLL diagnosis using ultrasound pictures. Furthermore, the limited sample sizes did not use standardized picture data and lacked external validation to assure the dependability of the results from the paper.

Wang et al. [10] in 2020 proposed another approach to classify focal liver lesions. During their study, they found that early detection is a more effective and helpful approach to diminish deaths caused by hepatic cancer. To obtain visual transition features of the liver four computer tomography (CT) scans are captured. In their study, they consider some of those characteristics on three levels. They extracted low-level features by direct images, mid-level were extracted by using Bag of Visual Words (BoVW), and high-level attributes by using a deep neural network. They suggested the K-CP method to understand tensor BoVW models for sparse representations of multi-phase medical images. Using the suggested strategy, they learned tensor codebooks and constructed BoVW models to extract spatial characteristics and temporal combinations of multi-phase healthcare pictures. Tensor-based methods often involve complex computations, especially when dealing with massive medical image datasets. The computational cost of learning tensor codebooks and building BoVW models may be high, requiring significant computational resources and time. In 2020, Yang et al. [11] proposed a convolutional neural network for FLL identification. They used US pictures in their proposal. Their DCNN-US model was built by combining clinical variables, ultrasonic characteristics, and radiomic signatures from liver and FLLs. The radiomic signature was established using a training cohort. To validate the performance of the produced model, internal validation (IV) and EV cohorts were employed. Same year, Lin et al. [12] proposed another method using a neural network with ResNet50 for detecting liver lesions. For their study they used ImageNet. For training and evaluation of the model, they used two datasets. One dataset has all tumors and the other one contains only the lesions were recognized. They split the dataset into a 7:2:1 ratio for training, validation, and testing groups. For this study, they use JPEG images and for their method, they use an application to crop images manually. This can be a reason why the model could identify wrong images. And as they use ResNet, which is not good for small and noisy datasets.

In 2021, Zhou et al. [3] developed an architecture for convolutional neural networks (CNNs) for the categorization of liver tumors in multi-phasic CT scans. The dataset or the CT images were gathered from a certain period of time using the picture archiving and communication system (PACS) in their organization. The researchers trained the data by the model which often requires specialized techniques to handle the complexity introduced by the different data sources or scales. Araújo et al. [1] in 2021 proposed algorithms which were convolutional neural networks (CNN), RetinaNet, and used U-Net to separate liver lesions in CT images. From the Liver Tumor Segmentation (LiTS), the dataset was assembled. RetinaNet does well with this suggested methodology when it comes to identifying large lesions, but it might have trouble with smaller lesions. There are a lot of factors in the U-Net technique, especially when employing deeper architectures or better-resolution photos. Thus, overfitting may result from this, especially in cases when the training dataset is tiny or undiversified. In 2021, Gavade et al. [13] proposed an algorithm named Adaptive Adam algorithm based DeepSegNet for identifying and grading hepatic tumors. The National Institutes of Health Clinical Center generated the DeepLesion dataset on November 27, 2020, and Ke Yan provided it. This dataset was used by the researchers for their proposed strategy. The CT scan pictures were used to compile this dataset to train the model. There are certain difficulties in recognizing PV sections out of 3D visuals with this approach. Additionally, handling small datasets and analyzing labeled data present challenges.

Survarachakan et al. [14] in 2022 developed four different types of methods for image-based liver analysis. The techniques mentioned were convolutional neural, autoencoders, deep belief, and generative adversarial networks. Two distinct datasets, 3D-IRCADb and MSD, which contain 20 abdominal CT scan pictures and 10 labeled datasets, respectively, were used to train the models. Due to the wealth of data available, the liver has been investigated mostly using MRI, US, and laparoscopic video since this research focuses on the analysis of image-based livers. Grouping and separation of the liver's different circulation systems In 2022, Dadoun et al. [15] proposed a network to identify, analyze, and characterize liver tumors in the liver on Ultrasound pictures. They implemented Detection Transformer (DETR) and Faster Region-Based Convolutional Neural Network (Faster R-CNN) as two distinct techniques in this work. The dataset known as 2551 B-mode abdominal ultrasound pictures, which was gathered between 2014 and 2018, was utilized to train the models. The focal liver lesion was detected with high specificity by the DETR approach; however, reduced specificity and accuracy were obtained due to the small test set size and low picture quality.

Wu et al. [16] established the AI multi-phase liver tumor segmentation (MULLET) approach for liver lesion identification. They used the Liver Diagnostic Initiative (AIELDI) dataset to train the model. This dataset, which is the massive multiphase set of data for the segmentation of hepatic lesions, includes a sum of 1229 CT images. The model they suggested has a better degree of accuracy for large lesions, but it has difficulty segmenting smaller lesions. Hille et al. [17] proposed the SWTR-UNet hybrid network with ResNet, transformer blocks, and UNet-style decoder for the detection of liver and hepatic lesions using MRI images in the same year. The liver lesion segmentation inter-observer variabilities demonstrated that the obtained segmentation accuracy was on par with expert segmentations that were done by hand. These models have trouble identifying smaller lesions. Kim et al. [18] in 2023 proposed two networks named 3D-CNN and CNN-LSTM for automatic classification between Hepatocellular carcinoma (HCC) with localized nodular hyperplasia. Together the ability to extract spatial data employing 2D-CNN, with the LTSM model makes it possible to acquire time-related connections among perfusion stages. Prakash et al. [4] proposed utilizing Convolutional Neural Networks (CNNs) based on DenseNEt to predict and categorize liver abnormalities. They use the liver CT scan pictures for the first step of the model's training. 10,000 samples of data were taken from Government General Hospital in Vijayawada for real-time training and testing. Previous models were unable to distinguish between lesions with sufficient clarity, while current models continue to struggle with accuracy.

In 2024 Zhou et al. [19] created a comprehensive deep learning method to categorize hepatic tumors as harmless or cancerous accurately. Incorporating the Contrast-enhanced ultrasound (CEUS) cycles into a 3D neural network architecture is the utmost logical style to use machine learning techniques for liver lesion identification with CEUS, as it maintains both spatial and temporal data. They suggested automatically using the differential classification of a liver tumor rooted in CEUS cycles using the CNN-LSTM model. They contrasted the suggested approach with a traditional machine learning-related time–intensity curve (TIC) and a 3D-CNN-based approach. Because 3D CNNs have more parameters than 2D CNNs, they can be more computationally expensive. Because LSTMs are susceptible to the vanishing gradient problem, particularly when working with lengthy sequences, this can cause training and inference to proceed more slowly and demand more computer power. This may make it more difficult for them to identify long-term dependencies in the data. Manjunath et al. [20] suggested a deep learning approach for identifying and categorizing liver diseases which they named Unit 60. They trained their model with a dataset that consists of liver disease CT images. The tumors with metastasis and cholangiocarcinoma were chosen and categorized using the extracted statistical, textural, and geometrical properties. In total, 864 liver cancer pictures were obtained using data augmentation from the publicly accessible 3Dircadb dataset. The deep computational model was developed and evaluated in the computed tomography visuals by changing the dicom pictures to another layout and eliminating the pictures without a tumor among the tagged dataset. The liver and its tumor each have their mask in the 3Dircadb dataset. This model is time-consuming and takes more time to train the model. Another thing is the number of CT images is not greater for investigation. Tangruangkiat et al. [21] proposed a pre-trained ResNet50 deep learning method in 2024. They collected 518 data for their dataset. The data they collected were B-mode ultrasound images. For this study, they employ a technique of validation to evaluate the precision and generality of graph-generative neural networks (GGN), ResNet18, ResNet50, and visual Geometry Group (VGG16). Poorer performance for accurately recognizing true negative cases and lower performance on a neural network-centered VGGNet included four types of FLLs (Cyst, HMG, HCC, and MLC) and two types of FLL. In 2024 Ying et al. [22] proposed a method called Liver Artificial Intelligence Diagnosis System (LiAIDS), which is established on a Convolutional neural network. All patients in their study underwent triple-phase CT scans, which involved non-contrast, arterial, and portal venous stages. There were three deep learning models were used in the development of LiAIDS to handle lesion categorization, segmentation of the liver, and lesion detection. Clinical data and contrast-enhanced CT scans were used to train and validate the system. Three primary modules comprised the structure of the suggested lesion identification, division, and categorization. The Faster R-CNN 21 two-stage detection technique served as the foundation for the lesion detection module's network design. The second stage used an R-CNN network to further classify FLL candidates and regress their locations. The two components of the lesion classification module were extracting characteristics from clinical and imaging sources. The attribute extractor backbone was made up of three distinct but separate networks that extracted image attributes of the venous, arterial, and non-contrast phases, correspondingly. In their study, they did not point out what privacy method they applied for the dataset. Large centralized datasets are needed for CNN training, which might cause privacy issues, particularly when handling private medical data. CNNs that were trained on such unbalanced data might make predictions that are prejudiced or struggle in minority classes. Lou et al. [23] present the SDR-Former, a novel architecture designed especially for 3D multi-phase CT and MR image classification of liver lesions. SDR-Former is a versatile framework meticulously crafted for the intricate task of 3D multi-phase CT/MR imaging liver lesion classification. To bolster SNN's display of features capability, a groundbreaking hybrid network dubbed DR-Former was ingeniously devised. Leveraging two diverse clinical datasets for validation—a comprehensive eight-phase MR dataset encompassing seven distinct lesion categories alongside a multi-phase CT scan data featuring two kinds of tumors—they meticulously scrutinized the efficacy of our SDR-Former approach. Moreover, it's imperative to acknowledge that the conventional level of imaging merging approach poses a potential concern of diminishing feature illustration, particularly across specific stages. These insightful findings underscore the potential limitations of relying solely on SNN in scenarios involving dual-phase circumstances. This emphasizes the necessity for innovative methodologies like DR-Former to navigate such complexities effectively.

2.2 *Privacy-Preserving Techniques*

Techniques for protecting the confidentiality and privacy of images are referred to as image privacy-preserving techniques. These technologies are extremely essential in the field of digital communication and broad image transfer, since photographs may include private or sensitive information that unauthorized parties should not see. Many privacy-preserving techniques have attracted a lot of attention in both industrial and academic domains. Table 2 shows a comparison of privacy-preserving techniques.

First of all, the Adversarial Privacy-preserving technique used in this [24] paper for image-based data. By tricking facial recognition algorithms, it successfully prevents harmful usage of publicly shared face photos. Depending on the particular attack scenario and the original sample's accessibility, the technique's efficacy may change.

Another privacy-preserving technique is Synthetic Data Generation which is used in [25]. To create synthetic data, the study used generative models such as variational autoencoders, Bayesian networks, and Hidden Markov models. Data usefulness can be maintained by using synthetic data, which can maintain statistical characteristics and correlations from the original data. There is some information loss because generative models only offer a lower-dimensional approximation of the actual data distribution.

Table 2 Observation of Privacy-preserving methods

Techniques	Accuracy	Privacy preserved	Utility	Information loss	Computational complexity	Communication complexity
Adversarial perturbation	Precise	Precise	Precise	Moderate	Precise	Minimal
Synthetic data generation	Moderate	Moderate	Precise	Minimal	Moderate	Minimal
De-Identification	Precise	Moderate	Precise	Minimal	Minimal	Minimal
Role-Based access control (RBAC)	Precise	Moderate	Precise	Minimal	Minimal	Minimal
Differential privacy	Minimal	Precise	Minimal	Minimal	Minimal	Minimal
Federated learning	Precise	Precise	Moderate	Minimal	Minimal	Minimal

The de-identification technique is used for CT scan pictures in DICOM format in [26]. This will preserve patient privacy by eliminating private information from medical photos. A laborious and intricate procedure that necessitates thorough assessment prior to implementation.

Another privacy-preserving technique is Role-Based Access Control (RBAC), to improves data security and privacy by guaranteeing that only authorized users can access healthcare data [27]. Yet role explosion can contribute to increased complexity, security risks, and decreased system efficiency in quickly evolving healthcare systems.

On the other hand, by adding noise to search results, a mathematical technique known as differential privacy is used to protect individual data. Clients are more inclined to divulge private details if they believe that this technique has decreased the potentiality of insights breaches [28].

In the above, there mentioned many privacy-preserving techniques but besides those techniques, the Federated Learning technique is used in this paper. This machine learning technique involves training models independently on various devices or servers, and then combining their gains to improve the model as a whole. This approach differs from centralized machine learning, which processes data by moving it to a central place. Compared to the other techniques which were mentioned earlier, it provides better privacy protection. In Table 2, there is a comparison shown between all of the techniques. In addition, it has less interaction complexity than both of the models and lower computational complexity than the Adversarial Perturbation, Synthetic Data Generation, and Differential Privacy. Hence, the authors chose Federated Learning as an image-processing paradigm to safeguard user privacy.

3 Proposed Methodology

This part will begin with a study of the dataset, including a detailed review of its components and characteristics. We will then discuss our suggested framework, defining its structure and important components.

3.1 Dataset Description

For both training and validation in this work, we utilized the dataset from the Liver and Liver Tumor Segmentation. 15 computed tomography pictures with 22 liver lesions for validation and 115 CT scans with 261 liver lesions for training made up the dataset. Because the cases in the dataset came from seven different institutions, there were differences in the CT scanners and methods that were employed.

- There were 261 liver lesions in 115 CT images that made up the training dataset.
- There were 22 liver lesions in 15 CT images that made up the validation dataset.
- The radial plane's picture resolution ranged from 0.56 up to 1.0 mm.
- Section counts for each scan ranged from 42 to 1026.
- With the average lifespan was 68 years (±10 years deviation from the standard), the test set comprised 10 female and 16 male patients.
- With an average of 4.0 (±2.6) lesions per patient, there were 105 lesions in total.
- The lesions were categorized into three sizes: smaller than 10 mm (n = 30), 10–20 mm (n = 35), and bigger than 20 mm (n = 40). The lesions ranged from 1.10 to 89.9 mm, with an average size of 19.4 mm (+–15.0 mm).

Three seasoned radiologists evaluated each case, which included hepatic metastasis masks segmented by various radiologists at each clinical site. A biobank that was registered with the Canadian Tissue Repository Network provided the cases, which were selected at random.

3.2 Proposed Framework

Federated learning in our system protects a user's privacy. In a classical machine learning structure, our system might have required screenshots to be submitted to a main server for it to be trained. Instead, it would have been placed on a single centralized server. This exposes the user's privacy and provides a tremendous opportunity for a security attack. However, federated learning allows neural network models to be constructed using dispersed collections of data rather than data that must be centered in one place, thus ensuring anonymity.

Federated learning enables the model to be cooperatively developed on all of these independent sources of data while keeping the data on the specific device or systems that provide them [29–31]. The model cannot be accessed by anyone on the individual machines or servers; only the most recent weights are communicated among each device and the centralized server [32–34]. This shows that sharing specific data sources with the central server, as well as knowing which data points the model is developing on, are optional. Federated learning enables protecting privacy machine learning since the model may be trained while the data remains secure and private on the devices or servers that give it.

Figure 1 illustrates the mechanics of Federated Learning (FL). FL operates with a central global model and multiple local servers connected to it. It comprises two main components: the central federated learning server with the global model, and the local data server with local federated learning models. Initially, the global model dispatches an initial update to the local servers. Each local model begins its task with this update, gathering data from the local server and adjusting accordingly.

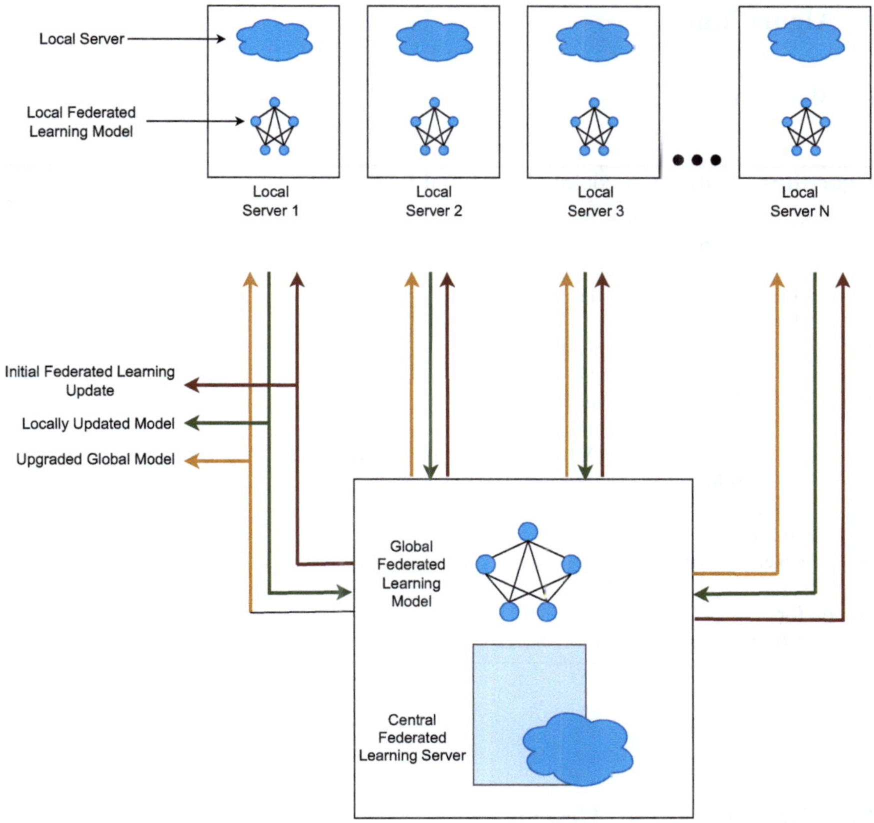

Fig. 1 Proposed framework for federated learning

Subsequently, each local model forwards its updated version to the general scheme. This general scheme combines all these local updates, integrating them to update itself comprehensively. Once updated, the global model disseminates its revised version to all local models. These local models incorporate the updates and resume their tasks, thus initiating a cyclic process of updates and iterations.

4 Algorithm

See Algorithm 1.

Algorithm 1 FedProx (Federated Proximal Learning)

Compute the global model W_{global} with W_0.
for $q = 1$ to Q **do**
    for $i = 1$ to N **in aligned do**
        Initialize the local model W_i as W_{global}.
        for $e = 1$ to E **do**
            Randomize the local dataset D_i.
            for each minibatch (x, y) in D_i **do**
                Compute the gradient $\nabla\ell(W_i; x, y)$.
                Update the local model:
                    $W_i \leftarrow W_i - \eta\left(\nabla\ell(W_i; x, y) + \mu(W_i - W_{global})\right)$
            end for
        end for
    end for
    Server aggregation:
    $W_{global} \leftarrow \frac{1}{N}\sum_{i=1}^{N} W_i$
end for
Output: The final global model W_{global}.

4.1 Explanation of Algorithm

In the given algorithm, initializing a global model, W_{global}, from W_0, is the first step in the FedProx method. Each client i initializes a local model W_i as a copy of W_{global} for each of the Q global iterations. Clients then use their randomized datasets to run E local training epochs. To keep W_i identical to W_{global}, the gradient is computed and utilized to update W_i with a proximal term at each epoch. The server obtains the final global model by aggregating the local models and updating W_{global}.

5 Result Analysis

The term "accuracy" is regarded as the most commonly used metric to evaluate the efficiency of a model. However, accuracy alone is insufficient to completely assess a model's flexibility, and it can occasionally be misleading (Abadi et al., 2021). As a result, the authors evaluated this show utilizing rightness and other measurements, counting F1-score, Precision, and Recall. To calculate these measurements, the variables such as False Negatives (FN), False Positives (FP), True Positives (TP), and True Negatives (TN) are considered.

A TP occasion in multi-class classification happens when the classifier precisely predicts that the test occurrence may be a part of lesson X. When a test illustration of lesson X is erroneously anticipated to be a diverse course Y, which comes about on an FN occasion. An FP occasion happens when a diverse course Y is erroneously anticipated as a part, of course, X. When a test protest from course Y is accurately anticipated to not have a place in lesson X, it is checked as TN. Equations (1)–(4) contains mathematical formulas for calculating various measurements.

$$Accuracy = \frac{TP + TN}{Total\ Number\ of\ Predictions} \tag{1}$$

$$Precision = \frac{TP}{TP + FP} \tag{2}$$

$$Recall = \frac{TP}{TP + FN} \tag{3}$$

$$F1\ Score = 2 \times \frac{Precision \times Recall}{Precision + Recall} \tag{4}$$

5.1 Performance Evaluation

In this subsection, the performance of the model we used on the dataset is shown with different criteria. Those criteria are metric analysis, loss and accuracy analysis, hyperparameter turning, complexity, and confusion matrix which are described in the following subsections.

5.1.1 Performance Metrics Analysis

We have evaluated many models like InceptionV3, VGG16, CNN, VGG19, and ResNet50. In these models, we get the highest performance in the model VGG19. The accuracy, precision, recall, and F1-score of the test and train data of this model are shown in Table 3. On the test dataset, the model's accuracy was 97.03%, which is the proportion of properly predicted cases relative to all instances. The percentage of accurate positive predictions among all positive predictions generated by the model is indicated by the precision of 81.05%. The model accurately detected 89.66% of all real positive events in the test data, with a recall of 89.66%. The model's total performance on the test data may be assessed using a single metric, the F1-score of 85.25%, which strikes an equilibrium between recall and precision.

Table 3 Performance metrics on test and train data

Federated model	Test data				Train data			
	Accuracy	Precision	Recall	F1-score	Accuracy	Precision	Recall	F1-score
FedVGG19	97.03%	81.05%	89.66%	85.25%	95.06%	76.50%	87.54%	81.65%

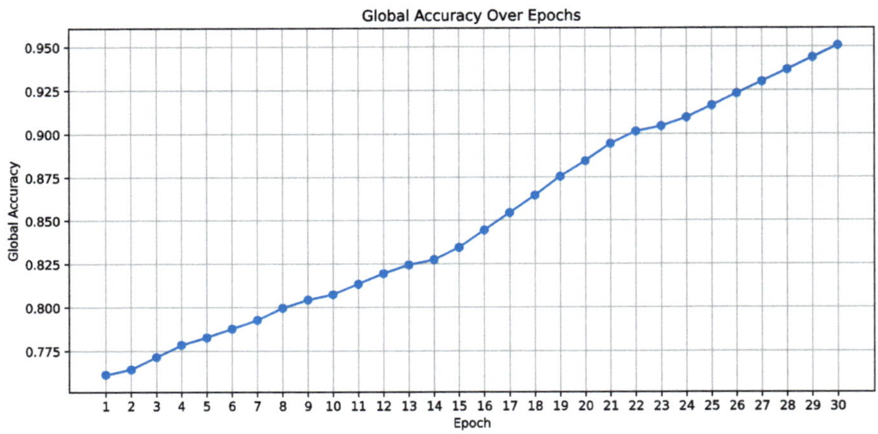

Fig. 2 Graph of accuracy in epochs

The model achieved 95.06% accuracy on the training dataset. The model's precision of 76.50% indicates the percentage of true positive predictions based on the training data. The model accurately identified 87.54% of positive occurrences in the training data, with a recall of 87.54%. The model performed well on this dataset, with an F1-score of 81.65% on train data.

5.1.2 Loss and Accuracy Analysis

Graphs of accuracy and loss are useful in machine learning research to evaluate model performance. Accuracy graphs show the percentage of correctly predicted values, whereas loss graphs reflect the difference between predicted and real values during training. Consider the trade-off between generalization and model complexity. Comparing loss and accuracy graphs can help guide model selection, hyperparameter adjustments, and model improvement strategies.

Figure 2 shows the accuracy scores of various machine learning models over different training epochs. The y-axis represents accuracy levels (0–1), while the x-axis shows epochs (1–24). The graph's colored lines indicate the accuracy trends for each model. The graph shows that the FedVGG19 model consistently outperforms other models, reaching near-perfect accuracy (close to 1) in later epochs. FedVGG19 has quite high accuracy ratings but varies significantly during training.

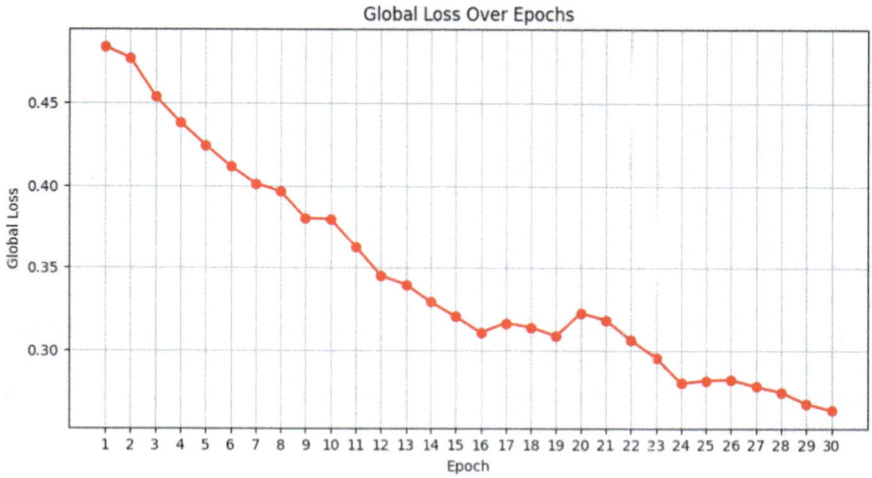

Fig. 3 Graph of loss in epochs

Figure 3 compares the loss values of machine learning models during training across epochs. The y-axis represents loss levels (0–1), while the x-axis shows epochs (1–30). The graph's lines, each colored differently to reflect a distinct model, show the loss trends for each model.

Overall, the graph demonstrates that the FedVGG19 model outperforms the other models with consistent accuracy, approaching near-perfect accuracy (close to 1) by the latter epochs. FedVGG19 exhibits comparatively good accuracy scores but with considerable variations over the training process. All models perform remarkably well in terms of accuracy, with the exception of FedVGG19.

5.1.3 Hyperparameter Tuning

In machine learning, choosing the best set of hyperparameters for a particular method or model is known as hyperparameter tuning. A variety of hyperparameters have been tweaked in order to get the best results for the models. The federated model name, learning rate, optimizer, epoch count, batch size, client count, dropout rate, activation functions, loss function, and image size are all listed in the following table.

In Table 4, it shows the hyperparameters of the federated model which we achieved during the training. Different hyperparameters include batch size, learning rate, number of epochs, client number, image size, and proximal coefficient.

The batch size is 128 for this model, which means the number of iterations used is 128. Learning rate is an important parameter and in this model, the learning rate is 0.001. The number of epochs is 30 to analyze the time needed to train this model. The training example number is 3 which refers to the number of clients and the proximal coefficient value is 0.5. The image size is 128,128. We have trained our model with

Table 4 Hyperparameters of model

Federated model	Batch size	Number of epochs	Number of clients	Learning rate	Image size	Proximal coefficient
FedVGG19	128	30	3	0.001	128X128	0.5

different image sizes as 255,256; 128,255, etc., yet we get the best results in this image size.

To improve the repeatability of the suggested approach, we included a detailed one-layer description of FedVGG19, as well as hyperparameters. This image provides a comprehensive understanding of the architecture, making it easy to replicate and apply the concept.

5.1.4 Computational Complexity

Training time is an important statistic in the field of computational complexity analysis and model performance comparison. A careful choice was made in our instance, where models were trained utilizing Google Colaboratory's openly accessible resources, to compromise between the quantity of epochs and resource constraints. To ensure an appropriate evaluation, we utilized all of the training time, which included User, Systime, and Wall time.

- User time is the amount of CPU time that a program or process uses in user mode to execute its instructions. It encapsulates the essential components of the computational burden of the application. This work's total CPU time is 5 h and 43 min.
- Contrarily, the system records the amount of CPU time utilized to run code in the kernel or system mode, including system calls and further authorized operations.
- Wall time, often known as actual time, precisely records the total amount of time that has passed, including execution time, system calls, input/output activities, and any possible delays. This work's total wall time is 56 min and 53 s.

Our findings show that models such as FedVGG19 have slightly longer training periods because of the limited picture size that they are designed to handle. It is crucial to note, however, that even the highest average training time clearly shows that our proposed approach is extremely well-suited to real-life situations.

5.1.5 Confusion Matrix

For evaluating the efficacy of classification algorithms, a critical machine learning tool is the confusion matrix, also known as the error matrix. It provides an extensive overview of the predictions made by a model given the actual data labels. The confusion matrix is an essential tool for assessing the efficacy and accuracy of the classification model since it shows the many kinds of mistakes the model generates, such as false positives, false negatives, true positives, and true negatives.

The confusion matrix is displayed in Figs. 4 and 5 provides insight into how various models fared when applied to the test set. FedVGG19's near-perfect accuracy in sample identification is proof positive of its outstanding performance. However, the other models also do exceptionally well in terms of classification accuracy, despite the stated deviations. The fact that FedVGG19 appears to perform worse than the other models is notable. These results show the accuracy with which FedVGG19 and the remaining classifiers categorize the test data; FedVGG19 is the model that performs the lowest out of all of them.

6 Conclusion and Future Work

Promising outcomes are observed with the suggested design for liver lesion detection. This essay covered tumor kinds and the medical history of the liver. The study provides an overview of the many methods for identifying liver lesions as well as the shortcomings of earlier methods. We employ a federated learning approach in this study to obtain more accurate results. Here, we use the FedProx algorithm to get the results. Specifically, we get an accuracy of 97.03% by using the FedVGG19

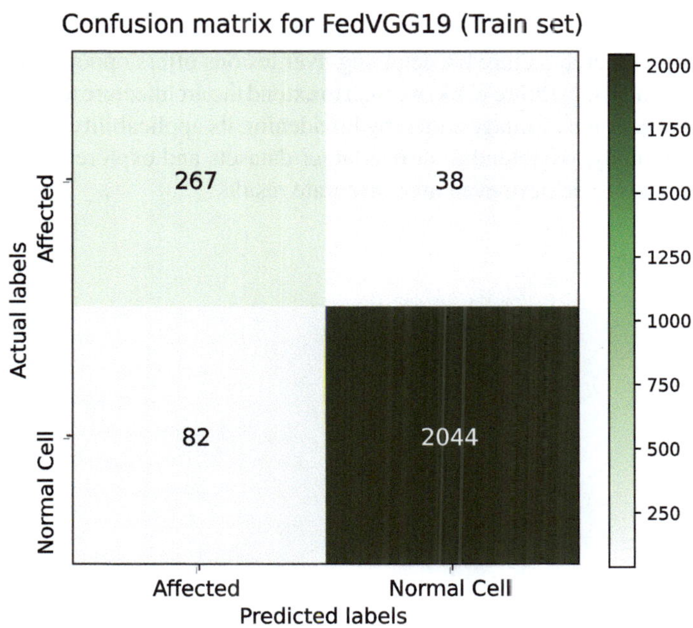

Fig. 4 Confusion matrix of trainset

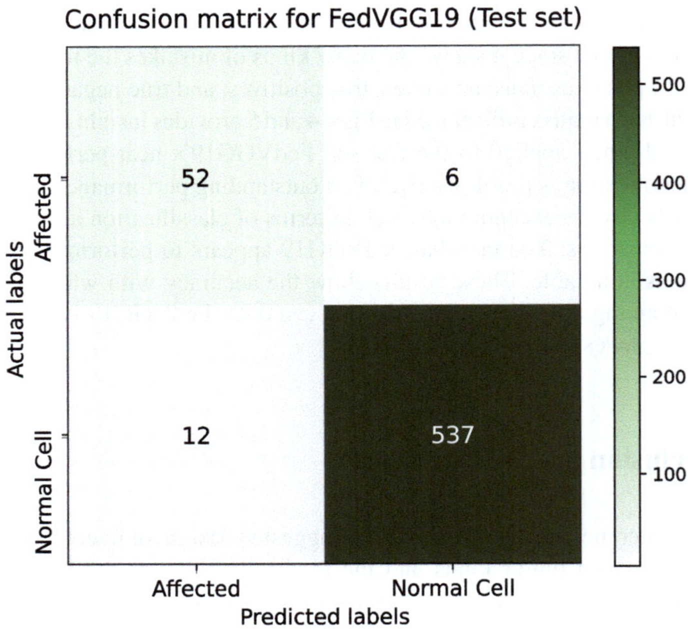

Fig. 5 Confusion matrix of testset

model in our dataset which contains a more diverse range of images to detect the liver lesion.

The proposed architecture for detecting liver lesions offers opportunities for further enhancements. In future work, we plan to extend the architecture to accommodate a more diverse range of images, thereby broadening its applicability in various contexts. Additionally, we intend to utilize larger datasets and explore other federated learning models to achieve even more accurate results.

References

1. J.D.L. Araújo, L.B. da Cruz, J.L. Ferreira, O.P. da Silva Neto, A.C. Silva, A.C. de Paiva, M. Gattass, An automatic method for segmentation of liver lesions in computed tomography images using deep neural networks. Expert. Syst. Appl. **180**, 115064 (2021)
2. M. Frid-Adar, E. Klang, M. Amitai, J. Goldberger, H. Greenspan, Synthetic data augmentation using GAN for improved liver lesion classification, in *2018 IEEE 15th International Symposium on Biomedical Imaging (ISBI 2018)*, (IEEE, 2018), pp. 289–293
3. J. Zhou, W. Wang, B. Lei, W. Ge, Y. Huang, L. Zhang, Y. Yan, D. Zhou, Y. Ding, J. Wu et al., Automatic detection and classification of focal liver lesions based on deep convolutional neural networks: a preliminary study. Front. Oncol. **10**, 581210 (2021)
4. N.N. Prakash, V. Rajesh, D.L. Namakhwa, S.D. Pande, S.H. Ahammad, A DenseNet CNN-based liver lesion prediction and classification for future medical diagnosis. Sci. Afr. **20**, e01629 (2023)
5. N. Shapira, J. Fokuhl, M. Schultheiß, S. Beck, F.K. Kopp, D. Pfeiffer, J. Dangelmaier, G. Pahn, A.P. Sauter, B. Renger et al., Liver lesion localisation and classification with convolutional neural networks: a comparison between conventional and spectral computed tomography. Biomed. Phys. & Eng. Express **6**(1), 015038 (2020)
6. H. Ying, X. Liu, M. Zhang, Y. Ren, S. Zhen, X. Wang, B. Liu, P. Hu, L. Duan, M. Cai et al., A multicenter clinical AI system study for detection and diagnosis of focal liver lesions. Nat. Commun. **15**(1), 1131 (2024)
7. W. Wang, Y. Iwamoto, X. Han, Y.-W. Chen, Q. Chen, D. Liang, L. Lin, H. Hu, Q. Zhang, Classification of focal liver lesions using deep learning with fine-tuning, in *Proceedings of the 2018 International Conference on Digital Medicine and Image Processing*, pp. 56–60 (2018)
8. E. Vorontsov, M. Cerny, P. Régnier, L. Di Jorio, C.J. Pal, R. Lapointe, F. Vandenbroucke-Menu, S. Turcotte, S. Kadoury, A. Tang, Deep learning for automated segmentation of liver lesions at CT in patients with colorectal cancer liver metastases. Radiol. Artif. Intell. **1**(2), 180014 (2019)
9. B. Schmauch, P. Herent, P. Jehanno, O. Dehaene, C. Saillard, C. Aubé, A. Luciani, N. Lassau, S. Jégou, Diagnosis of focal liver lesions from ultrasound using deep learning. Diagn. Interv. Imaging **100**(4), 227–233 (2019)
10. J. Wang, J. Li, X.-H. Han, L. Lin, H. Hu, Y. Xu, Q. Chen, Y. Iwamoto, Y.-W. Chen, Tensor-based sparse representations of multi-phase medical images for classification of focal liver lesions. Pattern Recognit. Lett. **130**, 207–215 (2020)
11. Q. Yang, J. Wei, X. Hao, D. Kong, X. Yu, T. Jiang, J. Xi, W. Cai, Y. Luo, X. Jing et al., Improving b-mode ultrasound diagnostic performance for focal liver lesions using deep learning: a multicentre study. EBioMedicine **56**, (2020)
12. I.L. Xi, J. Wu, J. Guan, P.J. Zhang, S.C. Horii, M.C. Soulen, Z. Zhang, H.X. Bai, Deep learning for differentiation of benign and malignant solid liver lesions on ultrasonography. Abdom. Radiol. **46**, 534–543 (2021)
13. A.B. Gavade, R.B. Nerli, S. Ghagane, Lesion detection and classification using sematic deep segmentation network, in *Smart Technologies in Data Science and Communication: Proceedings of SMART-DSC 2021*, (Springer, 2021), pp. 321–336
14. S. Survarachakan, P.J.R. Prasad, R. Naseem, J.P. de Frutos, R.P. Kumar, T. Langø, F.A. Cheikh, O.J. Elle, F. Lindseth, Deep learning for image-based liver analysis–a comprehensive review focusing on malignant lesions. Artif. Intell. Med. **130**, 102331 (2022)
15. H. Dadoun, A.-L. Rousseau, E. de Kerviler, J.-M. Correas, A.-M. Tissier, F. Joujou, S. Bodard, K. Khezzane, C. de Margerie-Mellon, H. Delingette et al., Deep learning for the detection, localization, and characterization of focal liver lesions on abdominal us images. Radiol. Artif. Intell. **4**(3), e210110 (2022)
16. L. Wu, H. Wang, Y. Chen, X. Zhang, T. Zhang, N. Shen, G. Tao, Z. Sun, Y. Ding, W. Wang et al., Beyond radiologist-level liver lesion detection on multi-phase contrast-enhanced CT images by deep learning. Iscience **26**(11), 108183 (2023)

17. G. Hille, S. Agrawal, P. Tummala, C. Wybranski, M. Pech, A. Surov, S. Saalfeld, Joint liver and hepatic lesion segmentation in MRI using a hybrid CNN with transformer layers. Comput. Methods Programs Biomed. **240**, 107647 (2023)
18. N. Kim, W.J. Lee, H.-J. Lee, Deep learning classification of focal liver lesions with contrast-enhanced ultrasound from arterial phase recordings, in *2023 International Conference on Electronics, Information, and Communication (ICEIC)*, (IEEE, 2023), pp. 1–3
19. H. Zhou, J. Ding, Y. Zhou, Y. Wang, L. Zhao, C.-C. Shih, J. Xu, J. Wang, L. Tong, Z. Chen et al., Malignancy diagnosis of liver lesion in contrast enhanced ultrasound using an end-to-end method based on deep learning. BMC Med. Imaging **24**(1), 68 (2024)
20. R. Manjunath, A. Ghanshala, K. Kwadiki, Deep learning algorithm performance evaluation in detection and classification of liver disease using CT images. Multimed. Tools Appl. **83**(1), 2773–2790 (2024)
21. S. Tangruangkiat, N. Chaiwongkot, C. Pamarapa, T. Rawangwong, A. Khunnarong, C. Chainarong, P. Sathapanawanthana, P. Hiranrat, R. Keerativittayayut, W. Sungkarat et al., Diagnosis of focal liver lesions from ultrasound images using a pretrained residual neural network. J. Appl. Clin. Med. Phys. **25**(1), e14210 (2024)
22. H. Ying, X. Liu, M. Zhang, Y. Ren, S. Zhen, X. Wang, B. Liu, P. Hu, L. Duan, M. Cai et al., A multicenter clinical AI system study for detection and diagnosis of focal liver lesions. Nat. Commun. **15**(1), 1131 (2024)
23. M. Lou, H. Ying, X. Liu, H.-Y. Zhou, Y. Zhang, Y. Yu, Sdr-former: a siamese dual-resolution transformer for liver lesion classification using 3d multi-phase imaging (2024). arXiv:2402.17246
24. J. Zhang, J. Sang, X. Zhao, X. Huang, Y. Sun, Y. Hu, Adversarial privacy-preserving filter, in *Proceedings of the 28th ACM International Conference on Multimedia*, pp. 1423–1431 (2020)
25. T. Stadler, B. Oprisanu, C. Troncoso, Synthetic data–anonymisation groundhog day, in *31st USENIX Security Symposium (USENIX Security 22)*, pp. 1451–1468 (2022)
26. A. Shahid, M.H. Bazargani, P. Banahan, B. Mac Namee, T. Kechadi, C. Treacy, G. Regan, P. MacMahon, A two-stage de-identification process for privacy-preserving medical image analysis, in *Healthcare*, vol. 10 (MDPI, 2022), p. 755
27. T. Garg, N. Kagalwalla, S. Puthran, P. Churi, A. Pawar, A novel approach of privacy-preserving data sharing system through data-tagging with role-based access control. World J. Eng. **20**(1), 12–28 (2023)
28. B.S. Diba, J.D. Plabon, M.M. Rahman, D. Mistry, A.K. Saha, M. Mridha, Explainable federated learning for privacy-preserving bangla sign language detection. Eng. Appl. Artif. Intell. **134**, 108657 (2024)
29. D. Mistry, M. Mridha, M. Safran, S. Alfarhood, A.K. Saha, D. Che, Privacy-preserving on-screen activity tracking and classification in e-learning using federated learning. IEEE Access (2023)
30. M.G. Kibriya, D. Mistry, D. Mistry, M.T. Rahman, S.H. Ovi, A. Tabassum, S. Hossain, Privacy-preserving vision-based detection of pox diseases using federated learning, in *Data-Driven Clinical Decision-Making Using Deep Learning in Imaging*, (Springer Nature Singapore, Singapore, 2024), pp. 105–122
31. A. Tabassum, S.H. Ovi, S. Hossain, M.R. Tonmoy, M.S.H. Shovon, M.R. Hussein, D. Mistry, Privacy preserving breast cancer prediction with mammography images using federated learning, in *Data-Driven Clinical Decision-Making Using Deep Learning in Imaging*, (Springer Nature Singapore, Singapore, 2024), pp. 227–245
32. D. Mistry, M.R. Tonmoy, M.S. Anower, A.T. Hasan, Federated transfer learning for vision-based fall detection, in *International Conference on Big Data, IoT and Machine Learning*, (Springer, 2023), pp. 961–975

33. D. Mistry, J.D. Plabon, B.S. Diba, S. Mukta, M.F. Mridha, Federated learning-based architecture for personalized next emoji prediction for social media comments. IEEE Access 1–1 (2024). https://doi.org/10.1109/ACCESS.2024.3448470
34. M.S.H. Shovon, Z. Sultana, J. Shin, M.A. Hamid, D. Mistry, Incorporating residual connections into a multi-channel CNN for lung cancer detection in digital pathology, in *Data-Driven Clinical Decision-Making Using Deep Learning in Imaging*. (Singapore, Springer Nature Singapore, 2024), p.189

Privacy-Preserving Kidney Stone Detection from X-Ray Image Using Federated Learning

Jayonto Dutta Plabon, Md. Mountasir Rahaman, Moshiur Rahman Tonmoy, Md. Golam Kibriya, and Adit Ishraq

Abstract Kidney stones are hardened masses that form in the kidneys and can block the urinary tract, causing severe symptoms. Imaging methods like X-rays, computed tomography (CT), and ultrasound are commonly employed for detection; however, their clinical interpretation may sometimes lack precision. Recent studies demonstrated the promise of artificial intelligence in kidney stone detection, yet many of them overlook important privacy concerns. To address this, we propose a novel approach utilizing federated learning (FL) for the detection of kidney stones in X-ray images. Our study seeks to demonstrate the effectiveness of FL by employing the Federated Averaging (FedAvg) technique in combination with the VGG16 model. FL enables model training collaboratively across decentralized data sources while preserving confidentiality. Our focus is on evaluating specific performance metrics, including accuracy, precision, recall, and F1-score, by comparing the results obtained from both test and training datasets distributed across multiple nodes. Remarkably, a 99.92% accuracy rate, 100% precision, 99.90% recall, and 99.95% F1-score is reached on FedAvg, in the test data and 99.98% accuracy and above for the other metrics on the train data. Besides this, Federated Learning has become the most successful way to keep the accuracy and generalization of the model through multiple decentralized datasets. These results reveal the tremendous power of Federated Learning to make the models work exceptionally well and generalize to out-of-distribution data.

J. D. Plabon · Md. M. Rahaman
Advanced Machine Intelligence Research Lab, Dhaka, Bangladesh
e-mail: 21101073@uap-bd.edu

M. R. Tonmoy (✉)
Department of Computer Science and Engineering, Bangladesh University of Business and Technology, Dhaka, Bangladesh
e-mail: moshiurtonmoy@bubt.edu.bd

Md. G. Kibriya
Department of EEE, Pabna University of Science and Technology, Pabna, Bangladesh

A. Ishraq
Computer and Systems Sciences, Stockholm University, Stockholm, Sweden
e-mail: adis6811@student.su.se

M. F. Mridha and N. Dey (eds.), *Federated Learning in Health Care Technology*, Studies in Computational Intelligence 1216, https://doi.org/10.1007/978-981-96-8353-6_4

Keywords Privacy preserving · Federated learning · Deep learning · Kidney stone detection

1 Introduction

Kidney stones are therefore solid concretions or calculi enclosed within the kidneys and originate from some of the mineral components present in the urine. The mentioned stones include calcium oxalate stones, uric acid, struvite, and cystine stones, and all differ in size, shape, and composition [1]. They can be the size of a grain of sand or big stones that may block the urinary system, resulting in signs illustrative of severe lower back pain, constant abdominal pain, blood-stained urine, nausea, vomiting, and fever with chill. Despite the fact that the majority of the kidney stones pass without complications, they are believed to cause a lot of discomfort and pain during their formation and transit in the wrong section of the system. The treatment options range from receiving medication for small stones and resize stone removal to surgeries for bigger and chronic stones [2].

The standard diagnostic tests include simple X-rays, CT scans, and ultrasounds, among others, in the diagnosis of kidney stones. Conventionally, understanding of these medical images is carried out by radiologists, but such analysis is tedious and very often includes errors resulting in wrong diagnoses and wrong timings. The use of deep learning in the medical field is recent; nevertheless, it displays a great promise of being able to provide qualitative analysis of medical images for detecting kidney stones [3]. However, training these models demands the accumulation of patient data, which if compromised can be disastrous; there could be increased cases data breaches. The protection of patient information is, on the other hand, a very sensitive area because this is also required by the law authorities [4].

Kidney stone introduction and analysis present certain problems in the sphere of medical diagnostics: high requirements for accuracy and anonymity [5]. Renal calculi or kidney stones diseases is a common ailment, and if not correctly diagnosed or if diagnosis is delayed, the patient can experience severe pain, infections, and even loss of kidney function [6]. The routine procedures for diagnosis have proven to work, but they are bound by equipment and skills' access, more so in the developing world. This gap emphasizes the current need for higher diagnostics for efficient and secure health care diagnosis across the various health care settings.

The trends such as the rising incidence of kidney stones and the sheer necessity to improve the standard of diagnosis, call for efficient technologies in order to implement the optimum diagnostic profile [7]. Kidney stones are rather common in the USA affecting approximately 1 in 11 people and the lifetime risk has been estimated as 19% in men and 9% in women. Given the high rate of kidney stones' recurrence, and the costs of associated complications, the diagnosis and management approaches play a significant role. The main rationale behind this study is to tackle the considerable privacy concerns when dealing with Personal Health Information in the context of developing machine learning classifiers for kidney stone detection [8]. Centralized machine learning solutions require gathering patient's data in a single database; therefore, it is relatively secure to personal data. Thus, one of the key challenges is to develop a mechanism for a decentralized approach that protects the patient's data from leakage while allowing multiple models to be trained collectively [9].

The current sets of methods of using machine learning for medical images could sometimes necessitate a combination of patient data from different sites under one database. Although the method is effective in training the model, it also raises many privacy and security concerns since the patient's information is sent and stored. In relation to the identification of kidney stones, the aforementioned concerns are exacerbated by the number of highly detailed images of patients' X-rays that are present [10]. In addition, the centralization leads to data heterogeneity issues because the X-ray image produced in one medical facility may have different quality and varying resolution compared to another one where it was produced; in addition, different diagnostic tests may be applied to the same X-ray image. Such variations are highly detrimental to the efficiency and mobility of the machine learning models.

The motivation for this study is therefore two-fold. First, the diagnostic technologies for kidney stones, which deal with high sensitivity and specificity as well as being compatible with other settings, are still lacking. Secondly, there is a need to ensure that the patient data is protected due to its very sensitive nature. The solution to this issue is the federation learning process, which allows institutions to train the machine learning models with the data collectively while at the same time not exposing the data, thus offering privacy and a better model resistance [11]. This paper introduces a novel approach for detecting kidney stones in X-ray images by applying the concept of federated learning combined with the FedAvg algorithm.

The key contributions of this research are outlined as follows:

- Proposed model that maintains confidentiality of the patient's information within the local institution, thus complementing strict data protection laws.
- To address the data heterogeneity issue, we presented a framework of the multiple-hospital collaborative training based on the proposed strategy called FedAvg for kidney stone identification.
- The models are evaluated and compared using metrics such as precision, recall, and F1-score across multiple X-ray datasets, including those representing kidney stones, cysts, healthy kidneys, and tumors.
- Using FedAvg and VGG16, the idea implied in the study is proven for kidney stone detection, its practical usability without the given precise accuracy rates.
- To mitigate the computational issues, we further show, in practice, how to deploy federated learning at scale and with efficiency in realistic healthcare contexts.

As a result, our framework has demonstrated that FL implemented together with modern machine learning approaches is a practical, private, and effective way to implement kidney stone detection that can be used in subsequent studies and applications. This research seeks to strengthen the possibilities of deploying privacy-preserving federated learning in medical imaging to improve diagnostic ability without the need to violate patients' rights to privacy [12]. The proposed framework deals not only with one of the most important needs, namely, the timely identification of kidney stones, but also meets tough standards of data privacy and security [13]. In this work, we aim to contribute to enhancing the security and effectiveness of AI-based diagnostic systems, enabling their widespread adoption in the healthcare sector and fostering positive impacts on patient health and systemic healthcare challenges. In this study, we introduce federated learning combined with advanced cryptography and differential privacy techniques as an optimal solution to balance data utility and privacy. Furthermore, we highlight the potential of federated learning in medical imaging, emphasizing the need for privacy-preserving, collaborative, and accurate diagnostic solutions.

2 Literature Review

The rise of machine learning has transformed medical image analysis, providing the potential for automated and objective detection of kidney stones. Many researchers have applied different approaches for accurate detection of kidney stones. This section reviews the various methods and approaches utilized in past studies for detecting kidney stones.

2.1 Recent Advances in Kidney Stone Detection Techniques

Yildirim et al. [6] used a heterogeneous modified artificial neural network (HMANN). Images are pre-processed to enhance contrasts before segmentation. Subsequently, three types of features are extracted: Haralick features, histogram features, and adaptive features. Among these, nine features—maximum probability, mean, skewness, kurtosis, cluster shade, homogeneity, sum average, and sum of squares—are selected for classification. The HMANN method for classification is structured into three architectures: Support Vector Machine (SVM), Multilayer Perceptron, and Backpropagation. Although they used an image dataset their provided citation contains a multivariate dataset with 400 instances and 24 features. They have reported the highest accuracy in kidney stone prediction using HMANN at 99.7%.

Ma et al. [5] developed a deep learning model using XResNet and tested it on a total of 1799 computed tomography (CT) images collected from Elazığ Fethi Sekin City Hospital, Turkey. From the total dataset, 790 images were identified as containing kidney stones, while 1009 images were classified as normal. The model achieved an accuracy of 96.82.

The research conducted by Ma et al. [5] has been expanded upon by Ruben and Narmadha [14]. While they claimed that the datasets were identical, the dataset descriptions provided do not align. Specifically, they reported 5077 images from normal subjects, 3709 from subjects with cysts, 1377 from subjects with kidney stones, and 2283 from subjects with tumors. They utilized the YOLOv7 [15] model for image segmentation and subsequently employed the Pulse Couple Neural Network (PCNN) [16] for classification purposes. Their improved model achieved an accuracy of 98.88.

Aksakall et al. [17] applied different ML techniques including Decision Trees (DT), Random Forest (RF), Support Vector Machines (SVM), Multilayer Perceptron (MLP), K-Nearest Neighbor (kNN), Naive Bayes (BernoulliNB), and convolutional neural networks (CNN). They utilized their dataset comprising 221 images, with 182 of them having kidney stones. The experiments showed that the Decision Tree Classifier (DT) produced the best classification results, achieving the highest F1-score of 85.3.

Cui et al. [3] developed a deep learning model using U-Net which was tested on 625 CT images. The model achieved a 93.86% accuracy in kidney stone detection.

Ahmed et al. [4] developed a transfer learning (TL) model leveraging a pre-trained VGG16, enhanced with explainable artificial intelligence (XAI), to detect kidney stones from X-ray images. The study used KUB X-ray images from the Department of Urology and Kidney Transplantation at MAYO Hospital in Lahore, Pakistan. The dataset consisted of 500 images from patients who underwent radiographic examinations for kidney stones between February 2021 and October 2022, with 250 images showing kidney stones and 250 considered normal. The model achieved a testing accuracy of 97.41.

2.2 Limitations of Existing Approaches and Research Insights

The mentioned studies showcase advancements in kidney stone detection using various machine learning techniques. However, they overlook the significant privacy concerns associated with handling medical data. The consolidation of sensitive patient data presents risks of unauthorized access and breaches. Ethical complexities surrounding patient consent, data ownership, and the risk of discrimination or stigmatization based on medical information further complicate data collection efforts.

To address privacy concerns, federated learning offers a practical solution. This approach enables model training to occur locally on individual devices that hold the data, eliminating the need for centralized data aggregation. Only model updates are exchanged between the central server and local devices, ensuring that sensitive medical data remains secure and localized. By utilizing federated learning, we can harness the collective dataset from decentralized devices while preserving patient privacy. This decentralized approach mitigates privacy risks, simplifies data collection, and fosters collaboration in medical research without compromising data security.

3 Privacy-Preserving Methodologies

Image privacy-preserving strategies are methods for preserving an image's privacy and confidentiality. In the age of electronic media and ubiquitous photo sharing, these tactics are especially crucial because images might include private or sensitive data that should be kept out of the hands of unauthorized people. A variety of privacy-preserving methods have gained considerable attention in both academic and industrial circles. Table 2 is the result of our comparison of several privacy-preserving strategies.

A cryptographic technique called homomorphic encryption preserves the security and privacy of encrypted data while allowing computations on it. However, real-world implementations face security obstacles, such as worries about circuit privacy and CCA security [18].

Perceptual information is concealed through block-wise image scrambling, and an adaptation network is proposed to identify these scrambled images. Typically, to mitigate the effects of photo encryption, two systems are needed: a classification system and a modification framework [19].

A mathematical technique called differential privacy adds noise to search results in order to secure private information. Users are more likely to divulge personal information if they believe that differential privacy has decreased the likelihood of data leaks [20].

ConvMixer is a neural network that combines self-attention techniques and convolutional layers to preserve image privacy while producing good results. It's an additional technique to protect the privacy of images [21]. In image classification tasks, though, it demonstrates vulnerability to adversarial attacks [22].

We decided to use FL as an alternative to the previously listed methodologies. This machine learning technique involves training models independently on various devices or servers, and then combining their gains to improve the model as a whole. This approach differs from centralized machine learning, which processes data by moving it to a central place. Federated Learning, which achieves a fair balance

Table 1 Comparison of privacy-preserving methodologies

Techniques	Data loss	Accuracy	Utility	Privacy preserved	Computational complexity	Communication complexity
Homomorphic encryption	Moderate	Advanced	Advanced	Moderate	Advanced	Advanced
Block-wise scrambled images	Inferior	Advanced	Advanced	Advanced	Inferior	Inferior
Differential privacy	Inferior	Inferior	Inferior	Advanced	Inferior	Inferior
ConvMixer	Moderate	Advanced	Advanced	Moderate	Inferior	Inferior
Federated learning	Inferior	Advanced	Moderate	Advanced	Inferior	Inferior

Table 2 Performance metrics of federated learning algorithms

Federated algorithm	Model	Test data				Train data			
		Accuracy	Precision	Recall	F1-Score	Accuracy	Precision	Recall	F1-Score
FedAvg	VGG16	99.92%	99.92%	99.92%	99.92%	99.98%	99.98%	99.98%	99.98%

between privacy and utility, is the most successful alternative for classifying photos, as we found after comparing different privacy-preserving algorithms (Table 1).

4 Methodology

4.1 Dataset Description

The dataset was collected from the PACS (Picture Archiving and Communication System) of several hospitals in Dhaka, Bangladesh, where patients had previously been diagnosed with kidney tumors, cysts, normal conditions, or kidney stones. The dataset was gathered using protocols for whole abdomen and urogram scans, with coronal and axial cuts selected from both contrast and non-contrast investigations. Each diagnosis was carefully selected from the DICOM research, and a batch of DICOM images was created for the relevant region associated with each radiological finding. These DICOM images were then converted to a lossless JPG format, with all patient data and metadata removed. After the conversion, each image was reviewed and confirmed by a radiologist and a medical technician to ensure the accuracy of the data.

Our dataset consists of 12,446 unique entries, including 3,709 cysts, 5,077 healthy kidneys, 1,377 kidney stones, and 2,283 tumors.

4.2 Machine Learning Model

A competition was used to introduce the CNN architecture VGG16 in 2014. The model consists of 16 layers, including 13 convolutional layers, 5 max-pooling layers, and 3 fully connected layers. In the convolutional layers of the network, 3×3 filters with a stride of 1 are used, and in the 2×2 max-pooling filters, a stride of 2. VGG16 is renowned for its simplicity because it has a manageable amount of hyperparameters. It can categorize 1000 different categories with an accuracy of 92.7% using about 138 million factors. Specifically designed for transfer learning, VGG16 can handle 224×224 input tensors with three RGB channels [23].

For a batch of N instances, the objective function (total loss) is the average cross-entropy loss over all instances in the batch:

$$\text{Loss} = \frac{1}{N}\sum_{j=1}^{N} L(\boldsymbol{y}^{(j)}, \hat{\boldsymbol{y}}^{(j)}) \tag{1}$$

In this formulation:

- $\boldsymbol{y}^{(j)}$ represents the true labels for the j-th instance.
- $\hat{\boldsymbol{y}}^{(j)}$ represents the predicted probabilities for the j-th instance.

4.3 Proposed Framework

Federated learning is a distributed machine learning paradigm that allows a group of clients, such as mobile devices or IoT devices, to collaboratively build a shared model while keeping the data local. Each client shares only the updates to the model with others, using its own local dataset to train the model. These updates are then aggregated to create the global model. During this procedure, clients get the global model, train on it locally, and send updates back. But it entails several communication cycles. It can be applied advantageously in such sectors as healthcare, banking, and individual mobile services because federated learning deals with such problems as data heterogeneity and privacy concerns. The general outline of federated learning is as follows:

$$\min_{\theta} \mathcal{L}(\eta; \theta) \text{ where } \mathcal{L}(\eta; \theta) = \sum_{t=1}^{T} \omega_t \mathcal{L}_t(\eta_t; \theta) \tag{2}$$

Here, $\mathcal{L}$ indicates the global loss function, acquired from a weighted combination of T local losses calculated from private data η_t and ω_t signifies the corresponding weight coefficients [24].

The use of FL for model training at mobile edge networks has the following benefits over traditional cloud-centric ML model training techniques [25]:

- Highly efficient bandwidth utilization: The amount of data sent to the cloud is minimized. For example, participating devices only transmit the updated model parameters for aggregation, rather than sending raw data for processing. This significantly reduces the load on backbone networks and lowers data transfer costs.
- Privacy: Users' raw data does not need to be uploaded to the cloud, enhancing privacy and reducing the risk of eavesdropping, assuming that federated learning participants and servers are trustworthy. Improved privacy will encourage more individuals to take part in collaborative model training, ultimately contributing to the development of more accurate inference models.
- Minimal latency: Federated learning allows machine learning models to be regularly trained and updated. At the same time, real-time local decisions, such as

event detection, can be made at edge nodes or end devices under the MEC (Multi-access Edge Computing) paradigm. As a result, latency is significantly reduced compared to making decisions in the cloud and sending them to the final devices. This is crucial for time-sensitive applications, such as self-driving car systems, where even the smallest delays could have catastrophic consequences [26–29].

4.4 Federated Averaging: FedAvg

Data heterogeneity is likely the most important issue, from a statistical perspective, in evaluating the efficacy of federated learning. It is likely that a single global model would result in a small IER_i for each i provided the data distribution Γ_i is approximately homogeneous across various clients. To solve this kind of problem, the federated averaging algorithm (FedAvg) was developed. This technique can be thought of as a localized version of stochastic gradient descent (SGD).

$$\min_{\omega} \frac{1}{\nu} \sum_{\iota \in [\mu]} \nu_\iota \Lambda_\iota(\omega, \Sigma_\iota) \tag{3}$$

Here, $\Lambda_\iota(\omega, \Sigma_\iota) := \sum_{\kappa \in [\nu_\iota]} \ell(\omega, \zeta_\kappa^{(\iota)})/\nu_\iota$ represents the $\iota^t h$ client's empirical risk minimization (ERM) objective, and $\nu = \nu_1 + \cdots + \nu_\mu$ specifies the total amount of training samples [30].

Description: Algorithm 1 illustrates FedAvg which is the core method behind federated learning. It consists of the local training of models at many clients and their subsequent aggregation to update the global model. Initially, the global model parameters w_{global} are set, and the learning rates for both global (α_g) and local (α_j) updates are defined. During each communication round t, a subset of clients $\mathcal{C}$ is randomly selected from the total client pool M. The server then broadcasts the current global model parameters w_{global} to these selected clients. Each client j initializes their local model w_j with the global parameters and performs local training for $\tau_{\max}$ iterations. In each local iteration, the client computes gradients $\text{grad}_j(w_j)$ from their local data and updates their model parameters w_j using their local learning rate α_j. After completing the local training, the client sends their model update $\Delta w_j = w_j - w_{\text{global}}$ back to the server. The server aggregates the received model updates by averaging them to compute $\Delta w_{\text{global}} = \frac{1}{|C|} \sum_{j \in C} \Delta w_j$, and updates the global model parameters w_{global} using the global learning rate α_g. This iterative process continues over multiple communication rounds, gradually refining the global model based on the aggregated updates from all participating clients, thus enabling collaborative learning without centralized data.

Algorithm 1 FedAvg

Initialize:
w_{global} : initial global model parameters
α_g : global learning rate
α_j : local learning rate for client j
T : total number of communication rounds
$\tau_{\max}$: number of local iterations per client
M : total number of clients
$\mathcal{C}$: subset of selected clients in each round
Δw_j : model update sent by client j
for each communication round $t = 1, \ldots, T$ do **do**
  select a subset of clients $\mathcal{C} \subseteq \{1, \ldots, M\}$
  broadcast global model parameters w_{global} to all selected clients $j \in \mathcal{C}$
  for each client $j \in \mathcal{C}$ in parallel do **do**
    initialize local model $w_j \leftarrow w_{\text{global}}$
    for each local iteration $\tau = 1, \ldots, \tau_{\max}$ do **do**
      compute local mini-batch gradient $\text{grad}_j(w_j)$
      update local model $w_j \leftarrow w_j - \alpha_j \text{grad}_j(w_j)$
    end for
    send the model update $\Delta w_j \leftarrow w_j - w_{\text{global}}$ to the server
  end for
  aggregate updates $\Delta w_{\text{global}} \leftarrow \frac{1}{|\mathcal{C}|} \sum_{j \in \mathcal{C}} \Delta w_j$
  update global model $w_{\text{global}} \leftarrow w_{\text{global}} + \alpha_g \Delta w_{\text{global}}$
end for

5 Result Analysis

In the following section, we will discuss the different types of metrics and the results obtained after training and testing the model.

5.1 Metrics Used for Evaluate The Result

The most commonly used criterion for assessing a model's effectiveness is known as "accuracy." However, it can be deceiving at times, and a model's accuracy alone does not indicate its flexibility. This model considers precision alongside other measures such as F1-score, Recall, and Accuracy. To calculate these metrics, the researchers have to consider the values False Negatives (FN), False Positives (FP), True Positives (TP), and True Negatives (TN).

In a classification with several classes, a "True Positive" event happens when the classifier correctly predicts that a test instance belongs to class X. A "False Negative" event happens when a test example of class X is incorrectly projected to be a different class Y. A "False Positive" occurrence occurs when another class Y is incorrectly projected as belonging to class X. "True Negative" indicates that a test instance from class Y is properly predicted to not belong to class X. We can see that Eqs. (4)–(7) contains mathematical equations for evaluating these measurements.

Accuracy is defined as the ratio of correct predictions made by a model.

$$\text{Accuracy} = \frac{\text{True Positive} + \text{True Negative}}{\text{Total Number Of Predictions}} \tag{4}$$

The degree of precision determines how many of a model's favorable predictions came true.

$$\text{Precision} = \frac{\text{True Positive}}{\text{True Positive} + \text{False Positive}} \tag{5}$$

Recall is a metric that measures a model's ability to discover and accurately anticipate every positive event in a dataset, hence reducing false negatives.

$$\text{Recall} = \frac{\text{True Positive}}{\text{True Positive} + \text{False Negative}} \tag{6}$$

The F1-score strikes a compromise between accurate and thorough categorization.

$$\text{F1-score} = 2 \times \frac{\text{Precision} \times \text{Recall}}{\text{Precision} + \text{Recall}} \tag{7}$$

5.2 *Experimental Setup*

The models were developed and trained using a research lab PC with the following specifications: an Intel i7 13th Gen CPU, 64GB RAM, 256GB SSD, 2TB HDD, and an RTX-4080 16GB graphics card. Training and implementation were carried out using the Python programming language, federated learning, and various popular machine learning frameworks, including Scikit-learn, TensorFlow, Keras, computer vision tools, and neural networks.

5.3 *Performance Evaluation*

Table 2 shows how the FedAVG version performed on the training and testing sets. This table summarizes essential evaluation measures, including accuracy, precision, recall, and F1-score for both training and testing sets. The FedAVG set of rules, utilizing the model VGG16, showed good results in both datasets, indicating strong performance metrics for kidney stone prediction tasks.

Table 2 presents a detailed comparison between the test and train data of the FedAVG algorithm with the VGG16 model. So if we look at the table, we can see that the test data obtained an accuracy, precision, recall, and F1-score of 99.92%. On

Table 3 Optimal values of hyperparameters

Federated algorithm	Model	Learning rate	Number of epochs	Batch size	Image size	Loss function	Optimizer
FedAVG	VGG16	0.001	20	64	128X128	Categorical Cross-entropy	Adam

the other side, the train data acquire an accuracy, precision, recall, and F1-score of 99.98%.

5.3.1 Hyperparameter Optimization

Hyperparameters are external configuration settings that are pre-set before training and are not learned from the data. These parameters influence how well a machine learning algorithm performs and directs the learning process. By finding the ideal configuration, hyperparameter tuning seeks to improve a model's performance by a methodical search for optimal hyperparameter values. Finding the set of hyperparameters that produces the most precise and broadly applicable model is the aim. We have optimized multiple hyperparameters to achieve optimal results in our models. The federate algorithm's name, model parameters, learning rate, epoch count, batch size, picture dimensions, loss function, and optimizer are among the details provided in Table 3. In this study, we selected several parameters to enhance the performance of federated models. The learning rate is a hyperparameter in machine learning algorithms that controls the step size during the optimization process. The number of iterations, or epochs, refers to how many times the learning algorithm processes the entire training dataset. Batch size is the number of data samples used in a single training iteration. "Image size" refers to the dimensions of an image, typically expressed in pixels as width and height. To train our models, we experimented with different sets of epochs, batch sizes, and image resolutions. After extensive testing, we found that a batch size of 64 and 20 epochs, with an image size of 128×128 pixels, provided the best performance.

Among the frameworks examined are Convolutional Neural Network (CNN), VGG16, VGG19, InceptionV3, MobileNetV2, and ResNet50. The aggregating method used to assess the frameworks was called Federated Averaging (FedAvg). Using the Federated Averaging (FedAvG) technique, VGG16 has obtained the highest test accuracy of 99.98% among the evaluated frameworks. It is notable because of this. Following closely behind with an accuracy of 99.57% is VGG 19. The CNN, VGG16, VGG19, InceptionV3, and MobileNetV2 frameworks all perform exceptionally well when using the Federated Averaging (FedAvg). Considering the fact that the accuracy scores were within allowable ranges, CNN and VGG19 produced outcomes that fell short of expectations. The efficiency measures show a continuous pattern that indicates the training dataset scores somewhat higher than the test dataset.

Other epoch counts, including 10 and 20 epochs, were also tried. We carefully tested 10, 15, and 20 epochs before settling on 20, had three clients, and finally selected a 256 by 256 picture size to achieve this precision. It is important to mention that we tried with picture sizes of 128X128 and 64X64. The fact that we followed the same procedure for all five models makes perfect sense. These models were implemented and trained on a research lab PC equipped with a `I7 13th Gen CPU, 64GB RAM, 256GB SSD, 2TB HDD`, and a `RTX-4080 16GB Graphics card`. Popular machine learning frameworks like `Scikit-learn` and `TensorFlow` were developed using the `Python` programming language.The fact that we followed the same procedure for all five models makes perfect sense. These models were implemented and trained on a research lab PC equipped with a `I7 13th Gen CPU, 64GB RAM, 256GB SSD, 2TB HDD`, and a `RTX-4080 16GB Graphics card`. Popular machine learning frameworks like `Scikit-learn` and `TensorFlow` were developed using the `Python` programming language.

5.3.2 Analysis of Accuracy and Loss

Two primary indicators utilized to assess the efficacy of the model are graphs displaying the frequency of errors and correct responses. Selecting models, adjusting hyperparameters, and enhancing machine learning performance in research environments all depend on the analysis of loss and accuracy graphs. These visual aids have a significant impact on model selection and improvement strategies. A measure called the accuracy graph shows the proportion of successfully identified cases to all instances. This is a very easy and intuitive technique to assess the effectiveness of the model.

Fig. 1 Global accuracy

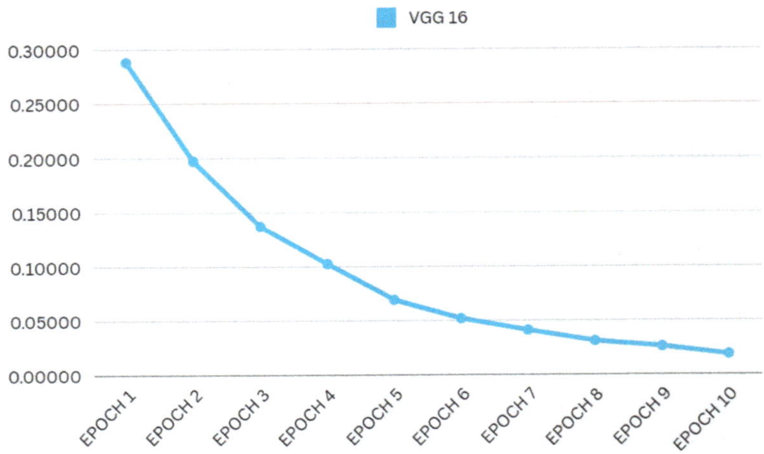

Fig. 2 Global loss

The graph in Fig. 1 illustrates the outstanding performance of the VGG16 model using the FedAvg approach. Notably, VGG16 consistently beats other models in terms of precision as training progresses, approaching nearly perfect scores (almost 1). Compared to VGG 19, CNN, and InceptionV3 perform wonderfully initially, but as the training epochs progress, they find it difficult to maintain high accuracy levels.

Figure 2 contrasts the machine learning model's loss values during training over epochs. The y-axis represents accuracy levels (0–1), and the x-axis represents epochs (1–10). The graph's lines, each colored differently to reflect a separate model, depict the accuracy trends for that model. The FedVGG16 model consistently outperforms other models, reaching near-perfect accuracy (close to 1) in later epochs. FedVGG16 has good accuracy scores but varies significantly during training. Except for Fed-VGG16, all models exhibit high accuracy.

5.3.3 Analysis of Confusion Matrices

An essential machine learning tool for evaluating the performance of classification models is the confusion matrix, often known as the error matrix. It provides a comprehensive summary of how a model's predictions compare to the actual labels on the data. The confusion matrix is an essential tool for assessing the performance and accuracy of the classification model since it shows the different kinds of errors that the model makes, such as false positives, false negatives, true positives, and true negatives [31].

For the test dataset, the model achieved 997 true positive classifications, accurately identifying instances with positive outcomes, while registering only 1 false positive, indicating a minor misclassification of negative instances as positive. Notably, there

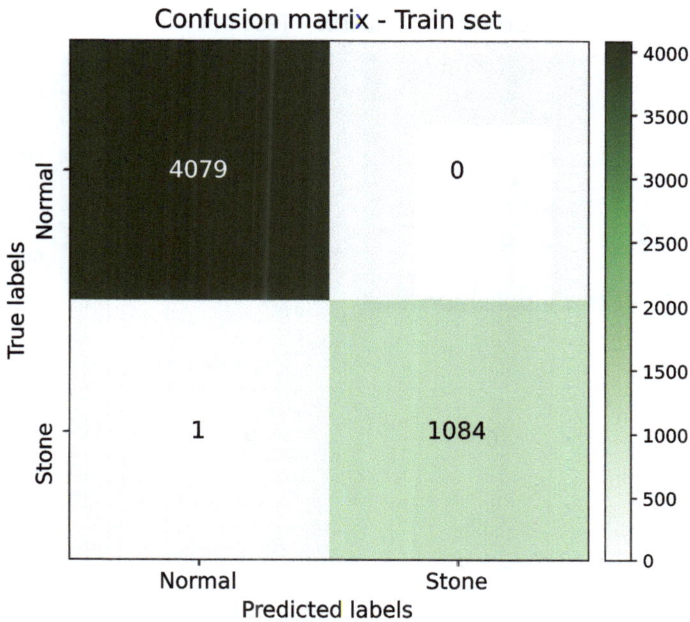

Fig. 3 Train confusion matrix of FedAvg-VGG16

were no true negative classifications, and 292 instances were incorrectly classified as negative when they were actually positive. This suggests a need for improved specificity in predicting negative outcomes (Fig. 3).

Figure 4 shows the test set of the confusion matrix. In contrast, the model performed well on the training dataset, with 4079 true positive classifications and only 1 false negative, suggesting excellent accuracy in identifying positive outcomes. However, it also showed 1084 false negatives, indicating difficulties in reliably predicting negative outcomes during training.

5.4 *Complexity Analysis*

Time and space complexity are additional computing resources that assess how effectively an algorithm performs in relation to data quantity. The difficulty of how to do such computations in a coordinated and effective manner is central to our work, which stems from the scattered nature of numerous equipment. In our case, where models were trained using Google Colaboratory's publicly available resources, we made a cautious decision to balance the number of epochs with resource constraints (Table 4).

The computational trials in our study revealed long processing durations, with a total CPU time of 1 day, 16 h, 56 min, and 5 s, and a total wall time of 7 h, 54 min,

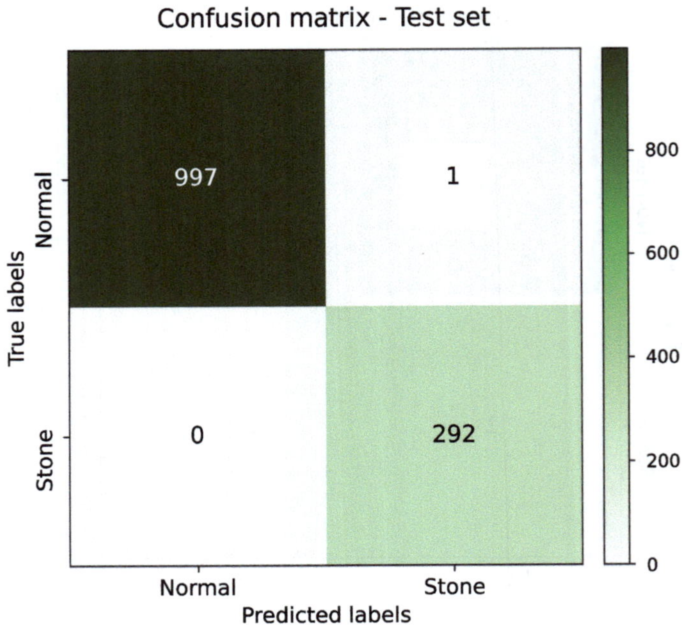

Fig. 4 Test confusion matrix of FedAvg-VGG16

Table 4 Total time taken by FedAvg with VGG16 model

Model	Total time taken	
	CPU time	Wall time
FedAvg-VGG16	1 day 16 h 56 min 5 s	7 h 54 min 36 s

and 36 s. Our approach, like the FedAVG algorithm, integrates an additional local optimization with a proximal term, resulting in larger computations depending on the optimization technique and the proxy term. One of the unique aspects of our system is that it operates on the user's devices.

6 Conclusions

Recent advancements in machine learning techniques have shown great potential in enhancing the accuracy and efficiency of kidney stone detection. However, the use of sensitive medical data raises significant privacy concerns. To address these issues, we have employed federated learning for kidney stone detection from X-ray images. Federated learning provides a decentralized approach to model training, enabling collaborative learning across multiple data sources while maintaining patient privacy. Our proposed framework not only improves the accuracy of kidney stone detection but also complies with stringent data security regulations and ethical standards. By mitigating privacy risks related to handling sensitive medical data, federated learning supports collaborative medical research without compromising patient privacy.

The experimental results demonstrate the effectiveness of federated learning, specifically using the FedAvg algorithm with the VGG16 model, attaining 99.92 test accuracy.

In future work, further refinements and optimizations of the federated learning framework can be explored to improve its scalability and efficiency. Additionally, extending the framework to incorporate other imaging modalities and clinical data sources can enhance the comprehensiveness of kidney stone detection systems.

References

1. F.L. Coe, A. Evan, E. Worcester et al., Kidney stone disease. J. Clin. Investig. **115**(10), 2598–2608 (2005)
2. T. Alelign, B. Petros, Kidney stone disease: an update on current concepts. Adv. Urol. **2018**(1), 3068365 (2018)
3. Y. Cui, Z. Sun, S. Ma, W. Liu, X. Wang, X. Zhang, X. Wang, Automatic detection and scoring of kidney stones on noncontrast CT images using STONE nephrolithometry: combined deep learning and thresholding methods. Mol. Imaging Biol. **23**, 436–445 (2021)
4. F. Ahmed, S. Abbas, A. Athar, T. Shahzad, W.A. Khan, M. Alharbi, M.A. Khan, A. Ahmed, Identification of kidney stones in KUB X-ray images using VGG16 empowered with explainable artificial intelligence. Sci. Rep. **14**(1), 6173 (2024)
5. F. Ma, T. Sun, L. Liu, H. Jing, Detection and diagnosis of chronic kidney disease using deep learning-based heterogeneous modified artificial neural network. Futur. Gener. Comput. Syst. **111**, 17–26 (2020)
6. K. Yildirim, P.G. Bozdag, M. Talo, O. Yildirim, M. Karabatak, U.R. Acharya, Deep learning model for automated kidney stone detection using coronal CT images. Comput. Biol. Med. **135**, 104569 (2021)
7. I. Aksakalli, S. Kadioğlu, Y.S. Hanay, Kidney x-ray images classification using machine learning and deep learning methods. Balk. J. Electr. Comput. Eng. **9**(2), 144–151 (2021)
8. U.S. Department of Health and Human Services, The HIPAA privacy rule (March 25 2024)
9. European Data Protection Supervisor, Health data in the workplace (March 25 2024)
10. B. McMahan, E. Moore, D. Ramage, S. Hampson, B.A. y Arcas, Communication-efficient learning of deep networks from decentralized data, in *Artificial Intelligence and Statistics*, (PMLR, 2017), pp. 1273–1282
11. D. Mistry, M.R. Tonmoy, M.S. Anower, A.S.M.T. Hasan, Federated transfer learning for vision-based fall detection, in *Proceedings of the 2nd International Conference on Big Data, IoT and Machine Learning*, ed. by M.S. Arefin, M.S. Kaiser, T. Bhuiyan, N. Dey, M. Mahmud (Springer Nature Singapore, Singapore, 2024), pp. 961–975
12. A. Mansoor, F. Naeem, M. Tariq, G. Kaddoum, Federated learning for privacy preservation in smart healthcare systems: a comprehensive survey. IEEE J. Biomed. Health Inform. **27**, 778–789 (2022)
13. N. Truong, K. Sun, S. Wang, F. Guitton, Y. Guo, Privacy preservation in federated learning: an insightful survey from the GDPR perspective. Comput. & Secur. **110**, 102402 (2021)
14. B. Reuben, C. Narmadha, Effective kidney stone prediction based on optimized yolov7 segmentation and deep learning classification. Int. J. Intell. Syst. Appl. Eng. **12**(1), 183–192 (2024)
15. C.-Y. Wang, A. Bochkovskiy, H.-Y.M. Liao, Yolov7: trainable bag-of-freebies sets new state-of-the-art for real-time object detectors (2022). arXiv:2207.02696Submitted

16. Z. Yi, J. Lian, Y. Qi, Z. Yu, H. Tang, Y. Ma, J. Liu, Deep pulse-coupled neural networks (2023). arXiv:2401.08649Submitted
17. I. Aksakalli, S. Kaçdioğlu, Y.S. Hanay, Kidney x-ray images classification using machine learning and deep learning methods. Balk. J. Electr. Comput. Eng. **9**(2), 144–151 (2021)
18. Z. Peng, Danger of using fully homomorphic encryption: a look at Microsoft seal (2019). arXiv:1906.07127
19. Z. Qi, A. MaungMaung, H. Kiya, Privacy-preserving image classification using convmixer with adaptative permutation matrix and block-wise scrambled image encryption. J. Imaging **9**(4), (2023)
20. R. Cummings, G. Kaptchuk, E.M. Redmiles, "I need a better description": an investigation into user expectations for differential privacy, in *Proceedings of the 2021 ACM SIGSAC Conference on Computer and Communications Security, CCS '21*, (Association for Computing Machinery, New York, NY, USA, 2021), pp. 3037–3052
21. S.T.A. Ozcelik, H. Uyanik, E. Deniz, A. Sengur, Automated hypertension detection using ConvMixer and spectrogram techniques with ballistocardiograph signals. Diagnostics **13**(2), (2023)
22. R. Iijima, M. Tanaka, I. Echizen, H. Kiya, On the adversarial transferability of convmixer models, in *2022 Asia-Pacific Signal and Information Processing Association Annual Summit and Conference (APSIPA ASC)*, pp. 1826–1830 (2022)
23. Z.N. Ashani, I.S.C. Ilias, K.Y. Ng, M.R.K. Ariffin, A.D. Jarno, N.Z. Zamri, Comparative analysis of deepfake image detection method using vgg16, vgg19 and resnet50. J. Adv. Res. Appl. Sci. Eng. Technol. **47**(1), 16–28 (2024)
24. N. Rieke, J. Hancox, W. Li, F. Milletari, H.R. Roth, S. Albarqouni, S. Bakas, M.N. Galtier, B.A. Landman, K. Maier-Hein et al., The future of digital health with federated learning. NPJ Digit. Med. **3**(1), 1–7 (2020)
25. W.Y.B. Lim, N.C. Luong, D.T. Hoang, Y. Jiao, Y.-C. Liang, Q. Yang, D. Niyato, C. Miao, Federated learning in mobile edge networks: a comprehensive survey. IEEE Commun. Surv. & Tutor. **22**(3), 2031–2063 (2020)
26. M.G. Kibriya, D. Mistry, D. Mistry, M.T. Rahman, S.H. Ovi, A. Tabassum, S. Hossain, Privacy-preserving vision-based detection of pox diseases using federated learning, in *Data-Driven Clinical Decision-Making Using Deep Learning in Imaging*, (Springer Nature Singapore, Singapore, 2024), pp. 105–122
27. A. Tabassum, S.H. Ovi, S. Hossain, M.R. Tonmoy, M.S.H. Shovon, M.R. Hussein, D. Mistry, Privacy preserving breast cancer prediction with mammography images using federated learning, in *Data-Driven Clinical Decision-Making Using Deep Learning in Imaging*, (Springer Nature Singapore, Singapore, 2024), pp. 227–245
28. D. Mistry, J.D. Plabon, B.S. Diba, S. Mukta, M.F. Mridha, Federated learning-based architecture for personalized next emoji prediction for social media comments. IEEE Access 1–1 (2024). https://doi.org/10.1109/ACCESS.2024.3448470
29. M.S.H. Shovon, Z. Sultana, J. Shin, M.A. Hamid, D. Mistry, Incorporating residual connections into a multi-channel CNN for lung cancer detection in digital pathology, in *Data-Driven Clinical Decision-Making Using Deep Learning in Imaging*. (Singapore, Springer Nature Singapore, 2024), p.189
30. S. Chen, Q. Zheng, Q. Long, W.J. Su, Minimax estimation for personalized federated learning: an alternative between FedAvg and local training? J. Mach. Learn. Res. **24**(262), 1–59 (2023)
31. D. Mistry, M.F. Mridha, M. Safran, S. Alfarhood, A.K. Saha, D. Che, Privacy-preserving on-screen activity tracking and classification in e-learning using federated learning. IEEE Access **11**, 79315–79329 (2023)

Federated Learning for Privacy-Preserving Healthcare Analytics: A Novel Framework for Fraud Detection in Healthcare

Md. Al Rafi, Gourab Nicholas Rodrigues, Md. Shahriar Mahmud Bhuiyan, Md. Nazmul Hossain Mir, and Md. Nahid Hasan

Abstract In the era of digital transformation, privacy-preserving solutions are paramount for enabling secure and collaborative healthcare analytics. This paper presents a Federated Learning (FL) framework for fraud detection in healthcare, ensuring privacy preservation and compliance with regulatory standards. The proposed approach integrates differential privacy and secure aggregation to protect sensitive patient and provider data while enabling multi-stakeholder collaboration. Evaluation on a comprehensive healthcare dataset demonstrates the effectiveness of the framework, achieving a testing accuracy of 93.5%, a precision of 94.2%, and an F1-score of 93.1%. The model's convergence is evident, with the global loss reducing from 1.345 in the initial round to 0.395 after 50 communication rounds. Furthermore, the impact of privacy budgets (ϵ) on performance was analyzed, showing that even with a highly stringent budget ($\epsilon = 0.5$), the framework maintains an accuracy of 93.0% and an F1-score of 91.2%. Communication efficiency optimizations reduce data exchange by up to 90%, demonstrating scalability across distributed healthcare systems. These results highlight the potential of the FL framework to transform healthcare analytics, ensuring privacy, accuracy, and efficiency in a multi-stakeholder environment. This study sets a foundation for deploying FL in real-world healthcare settings, addressing critical challenges in data security and collaborative analytics.

Keywords Federated learning · Privacy-preserving analytics · Healthcare fraud detection · Business intelligence · Differential privacy · Distributed machine learning · Secure aggregation

---

Md. Al Rafi · G. N. Rodrigues · Md. S. M. Bhuiyan · Md. N. H. Mir
Washington University of Science and Technology, Alexandria, VA, USA

Md. N. Hasan (✉)
University of Wisconsin-Milwaukee, Milwaukee, WI, USA
e-mail: hasan26@uwm.edu

M. F. Mridha and N. Dey (eds.), *Federated Learning in Health Care Technology*, Studies in Computational Intelligence 1216, https://doi.org/10.1007/978-981-96-8353-6_5

1 Introduction

Rapid digital transformation in the healthcare sector has led to a significant increase in the generation and utilization of healthcare data. From patient demographics and clinical histories to claim records and reimbursement details, these datasets have the potential to unlock valuable insights for fraud detection, personalized healthcare, and resource optimization [1]. However, sharing and centralizing such sensitive data pose critical privacy and security concerns, exacerbated by stringent regulations like HIPAA and GDPR [2, 3]. These challenges necessitate innovative solutions to enable collaborative analytics without compromising privacy.

Federated Learning (FL) is a promising paradigm to address these issues. FL enables multiple entities, such as hospitals and insurance companies, to train machine learning models collaboratively without sharing raw data [4]. By keeping data decentralized, FL ensures privacy preservation while leveraging the collective intelligence of distributed datasets. The application of FL in healthcare is particularly appealing, as it aligns with ethical and regulatory requirements while unlocking the potential for enhanced fraud detection and operational efficiency [5].

Healthcare fraud is a pervasive issue, costing the global healthcare system billions of dollars annually [6, 7]. Fraudulent claims, ranging from falsified services to inflated reimbursements, not only drain financial resources but also undermine trust among stakeholders. Traditional fraud detection approaches rely on centralized machine learning models, which require aggregating data from multiple sources. Such centralization is fraught with challenges, including data breaches, regulatory non-compliance, and stakeholder reluctance to share proprietary data.

Existing privacy-preserving methods, such as anonymization and encryption, are often insufficient for large-scale collaborative analytics. Anonymization can lead to information loss, while encryption significantly increases computational overhead [8, 9]. This creates a pressing need for scalable, privacy-preserving, and collaborative approaches that balance data utility and security.

To address these challenges, this paper proposes a Federated Learning-based framework tailored for privacy-preserving fraud detection in healthcare. The key contributions of this work are as follows:

- **Novel FL Framework**: We design an FL framework that integrates differential privacy and secure aggregation to enable collaborative fraud detection without sharing sensitive data.
- **Comprehensive Evaluation**: The framework is extensively evaluated on a large healthcare dataset, achieving a testing accuracy of 93.5% and an F1-score of 93.1%, demonstrating its efficacy in detecting fraudulent claims.
- **Privacy-Performance Trade-off Analysis**: We analyze the impact of privacy budgets (ϵ) on model performance, showing that even under stringent privacy constraints ($\epsilon = 0.5$), the framework maintains high accuracy (93.0%) and F1-score (91.2%).
- **Communication Efficiency**: To address scalability, the framework incorporates communication optimization techniques, reducing data exchange by up to 90%.

- **Real-World Applicability**: The proposed framework adheres to regulatory requirements, making it viable for deployment in real-world healthcare settings.

The remainder of this paper is structured as follows. Section 2 details the materials and methods, including data preprocessing, FL framework design, and privacy mechanisms. Section 3 presents the experimental setup and results, covering performance metrics, privacy analysis, and communication efficiency. Section 4 discusses the implications of the findings and the challenges of deploying FL in healthcare. Finally, Sect. 5 concludes the paper and outlines directions for future work.

2 Related Works

The application of Federated Learning in privacy-preserving healthcare analytics has garnered significant attention in recent years. This section reviews the existing literature on FL frameworks, privacy-preserving techniques, and fraud detection methodologies relevant to the proposed work.

Federated Learning has emerged as a transformative approach for addressing privacy concerns in healthcare applications while enabling collaborative data analysis. Sharma and Guleria [10] provide a comprehensive review of FL's role in disease detection, discussing its ability to preserve patient data confidentiality while improving model accuracy through distributed architectures. They explore various FL techniques, including horizontal, vertical, and federated transfer learning, alongside aggregation methods and challenges such as handling non-IID data and ensuring robust model performance. Chaddad et al. [11] delve into the application of FL in healthcare, emphasizing its utility in privacy-preserving tasks like medical image analysis and remote health monitoring. They highlight FL's ability to handle unbalanced data distributions and protect sensitive patient information under stringent data privacy regulations. Furthermore, the IEEE Internet of Things Journal outlines essential FL steps, including local training, parameter aggregation, and iterative updates, to create global models without centralized data sharing. Another detailed exploration by Li et al. [12] examines the integration of FL in clinical and operational settings, showcasing its potential to unify disparate datasets while maintaining compliance with data governance standards. Collectively, these studies underscore FL's pivotal role in modernizing healthcare by balancing innovation with privacy preservation.

The exploration of privacy-preserving federated learning in healthcare has gained significant traction, as evidenced by recent research contributions. Haque et al. [13] introduces a three-tier framework that combines blockchain technology, elliptic curve cryptography, and differential privacy. The architecture ensures secure data storage, authentication, and privacy-preserving model training. Through extensive experiments, the framework demonstrated higher accuracy levels of 97%, 83%, and 74% for breast cancer, heart disease, and diabetes datasets, respectively, compared to existing methods. However, the reliance on computationally intensive cryptographic operations may pose challenges for scalability and real-time applications. Similarly,

Yazdinejad et al. [14] leverages Trusted Execution Environments (TEEs) and Active Personalized Federated Learning (ActPerFL) to address privacy leakage and non-independent and identically distributed (non-IID) data challenges in healthcare FL settings. The model introduces an auditing mechanism to ensure process integrity and fairness. ActPerFL and Batch Normalization are used to enhance local and global model synergy, achieving superior performance in federated setups. Despite these advancements, the framework's reliance on TEE and its overhead for auditing mechanisms could limit its adoption in resource-constrained environments. In another study, Cao et al. [15] examines the role of FL in decentralizing sensitive medical data analysis. The proposed framework addresses issues of data heterogeneity and privacy through innovative aggregation techniques. While the study highlighted improvements in training efficiency and privacy safeguards, it also pointed out the limitations of existing models in handling real-world non-IID data distributions and scalability constraints.

Recent advancements in fraud detection have leveraged multivariate analysis to tackle healthcare insurance fraud. Settipalli and Gangadharan proposed the Weighted MultiTree Density-Based Clustering (WMTDBC) framework, an unsupervised approach to analyze healthcare claims data by constructing hierarchical Weighted MultiTree structures for categorical data and applying density-based clustering for continuous data [16, 17]. This method demonstrated superior detection capabilities compared to state-of-the-art models when applied to CMS Part B claims data, showcasing its strength in uncovering fraudulent patterns. However, its dependency on domain-specific thresholds and limited scalability across heterogeneous datasets restricts its broader applicability. In the domain of blockchain-integrated healthcare systems, Mohammed et al. introduced a machine learning-based framework for fraud detection in blockchain networks [18]. Utilizing six machine learning algorithms, including Random Forest, the framework excelled in identifying abnormal medical data and fraudulent transactions, achieving high accuracy and robustness against blockchain-specific attacks. While the results indicated significant improvements in security and scalability, the study highlighted limitations in handling diverse data streams and maintaining energy efficiency within blockchain operations.

While FL has shown promise in healthcare analytics, its application in fraud detection remains underexplored. Existing privacy-preserving techniques, though effective, often suffer from trade-offs in utility and scalability. This study addresses these gaps by proposing a scalable FL framework for fraud detection, incorporating advanced privacy mechanisms, and optimizing communication efficiency for real-world applicability.

3 Methods and Materials

The proposed methodology leverages Federated Learning to enable collaborative fraud detection in healthcare while ensuring privacy preservation across distributed datasets. The framework comprises several key components: data preprocessing, local model training, global model aggregation, and privacy-preserving mechanisms. Each stakeholder (e.g., hospitals and insurance companies) trains a local model on its private data, and only encrypted model updates are shared with a central server, avoiding raw data transfer. The global model is iteratively updated through weighted aggregation of local updates, balancing contributions based on dataset sizes. Differential privacy is integrated to protect individual data points by adding controlled noise to gradients, while secure aggregation ensures that model updates remain confidential. Communication efficiency is optimized through gradient sparsification and compression techniques, reducing the bandwidth required for multi-round training. Figure 1 presents the proposed architecture.

3.1 Data Preprocessing

The proposed methodology utilizes a comprehensive healthcare claims dataset that includes information on inpatient, outpatient, and beneficiary records. The preprocessing phase is critical for ensuring data quality, reducing noise, and enabling effective model training. The steps are outlined as follows:

Handling Missing Values The dataset contains several missing values, particularly in fields related to diagnosis and procedure codes. To address this:

$$\text{Imputation: } x_i = \begin{cases} \mu_j, & \text{if } x_i \text{ is missing in feature } j, \\ 0, & \text{if feature } j \text{ is binary.} \end{cases} \quad (1)$$

where x_i represents a missing value in feature j, and μ_j is the mean of the feature.

Normalization Continuous features such as reimbursement amounts were normalized to improve model convergence:

$$\tilde{x}_i = \frac{x_i - \min(x)}{\max(x) - \min(x)}. \quad (2)$$

This ensures that all features are scaled to the range [0, 1].

Feature Engineering New features were derived to enhance the dataset:

- **AdmitForDays**: Calculated as the difference between discharge and admission dates:

$$\text{AdmitForDays}_i = \text{DischargeDt}_i - \text{AdmissionDt}_i + 1. \quad (3)$$

- **Age**: Derived from the beneficiary's date of birth (DOB) and the dataset's reference year (2009):

$$\text{Age}_i = \text{Year}_{\text{ref}} - \text{Year}_{\text{DOB}_i}. \quad (4)$$

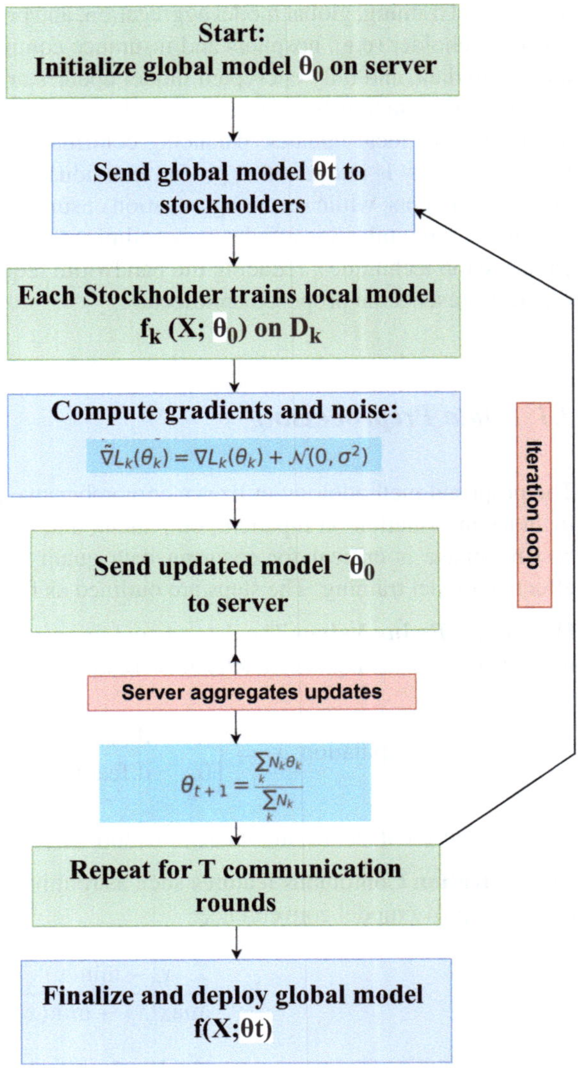

Fig. 1 The proposed architecture

– **WhetherDead**: Binary indicator of mortality:

$$\text{WhetherDead}_i = \begin{cases} 1, & \text{if DOD}_i \neq \text{NaN}, \\ 0, & \text{otherwise.} \end{cases} \tag{5}$$

Label Encoding and Categorical Features Categorical variables such as provider IDs and physician details were converted into numerical representations using one-hot encoding.

3.2 Proposed Methodology

The proposed methodology integrates Federated Learning (FL) with privacy-preserving mechanisms to enable secure and collaborative healthcare analytics. The workflow consists of several key components: data partitioning, model design, distributed training, and privacy enhancements. Each component is elaborated below.

Data Partitioning and Distribution The healthcare data is distributed across multiple stakeholders, such as hospitals and insurance companies, with each stakeholder holding a private dataset D_k. The datasets contain heterogeneous features, including patient demographics, claim details, diagnosis codes, and reimbursement amounts. The local data distributions are defined as

$$D_k = \{(x_i, y_i) \mid x_i \in \mathbb{R}^n, y_i \in \{0, 1\}, i = 1, 2, \ldots, N_k\},$$

where x_i represents the feature vector, y_i denotes the target label (fraudulent or not), N_k is the number of samples at stakeholder k, and K is the total number of stakeholders.

Model Architecture Design The global model is a deep neural network with the following layers:

- **Input Layer**: Takes the feature vector x, including diagnosis codes, procedure codes, reimbursement amounts, and demographic attributes.
- **Hidden Layers**: Employs multiple fully connected layers with ReLU activation:

$$h^{(l)} = \max(0, W^{(l)} h^{(l-1)} + b^{(l)}),$$

  where $h^{(l)}$ represents the output of layer l, $W^{(l)}$ and $b^{(l)}$ are the weights and biases of the layer, respectively.
- **Output Layer**: A single neuron with a sigmoid activation function for binary classification:

$$\hat{y} = \sigma(W^{(L)} h^{(L-1)} + b^{(L)}),$$

  where $\sigma(z) = \frac{1}{1+e^{-z}}$ is the sigmoid function.

The model parameters $\theta = \{W, b\}$ are initialized randomly and updated iteratively during the training process.

Distributed Training Process The training process in the FL framework involves the following steps:

Step 1: Local Model Training Each stakeholder trains a local model $f_k(x; \theta_k)$ using its private dataset D_k. The local training minimizes the empirical loss:

$$L_k(\theta_k) = \frac{1}{N_k} \sum_{i=1}^{N_k} \ell(f_k(x_i; \theta_k), y_i),$$

where $\ell(\cdot)$ is the binary cross-entropy loss:

$$\ell(\hat{y}_i, y_i) = -y_i \log(\hat{y}_i) - (1 - y_i) \log(1 - \hat{y}_i).$$

Step 2: Model Aggregation The local models are sent to a central server, which performs a weighted aggregation of parameters:

$$\theta = \frac{\sum_{k=1}^{K} N_k \theta_k}{\sum_{k=1}^{K} N_k}.$$

This step ensures that stakeholders with larger datasets have a greater influence on the global model.

Step 3: Global Model Update The updated global model $f(x; \theta)$ is sent back to all stakeholders for further local training. This process is repeated for T communication rounds until convergence.

Privacy-Preserving Mechanisms To ensure robust privacy, the methodology incorporates the following mechanisms:
Differential Privacy Noise is added to local gradients before sharing them with the server:

$$\tilde{\nabla} L_k(\theta_k) = \nabla L_k(\theta_k) + \mathcal{N}(0, \sigma^2),$$

where $\mathcal{N}(0, \sigma^2)$ is Gaussian noise with variance σ^2.

Secure Aggregation To prevent adversaries from reconstructing individual updates, secure multi-party computation (SMPC) is employed. The server receives only encrypted updates:

$$\text{Enc}(\theta_k) = \theta_k + r_k,$$

where r_k is a random noise vector shared among stakeholders.

Convergence Analysis The global optimization objective is to minimize the overall loss:

$$L(\theta) = \frac{1}{\sum_{k=1}^{K} N_k} \sum_{k=1}^{K} \sum_{i=1}^{N_k} \ell(f(x_i; \theta), y_i).$$

Using the aggregated updates, the convergence rate of the model is bounded by

$$L(\theta^{(t+1)}) - L(\theta^{(t)}) \leq -\eta |\nabla L(\theta^{(t)})|^2 + C\eta^2,$$

where η is the learning rate and C is a constant representing the variance of updates.

3.3 Evaluation and Analytics

To assess the effectiveness of the FL framework, several experiments were conducted:

- **Model Performance**: Metrics such as accuracy, precision, recall, and F1-score are computed to evaluate fraud detection capabilities.
- **Privacy Analysis**: The level of privacy preservation is quantified using privacy budgets ϵ from differential privacy.
- **Communication Overhead**: The number of bytes exchanged per communication round is analyzed.

4 Results

This section presents the evaluation results of the proposed Federated Learning (FL) framework for privacy-preserving healthcare analytics. The results include performance metrics, privacy analysis, communication efficiency, and comparisons with baseline methods.

4.1 Dataset Description

The dataset used in this study is a comprehensive healthcare claims dataset comprising information from multiple stakeholders, including hospitals, insurance companies, and healthcare providers [19]. The data is divided into training and testing subsets, containing records of beneficiaries, inpatient and outpatient claims, and associated labels indicating potential fraud. The dataset's hierarchical structure facilitates the integration of various features for fraud detection while preserving privacy.

- **Train Data**: Contains 5410 provider records with binary labels (*Yes* for potential fraud and *No* for non-fraudulent providers).
- **Test Data**: Includes 1353 provider records for evaluation, without fraud labels.
- **Beneficiary Data**: Includes demographic and health condition attributes of 138,556 beneficiaries in the training set and 63,968 in the test set.
- **Inpatient and Outpatient Claims**: Covers 40,474 inpatient and 517,737 outpatient claims in the training set and 9551 inpatient and 125,841 outpatient claims in the test set.

Table 1 provides an overview of the dataset statistics across its various components.

The dataset includes diverse features such as demographic attributes (e.g., age, gender, race), clinical conditions (e.g., chronic diseases), and financial information (e.g., reimbursement amounts, deductible amounts). This rich set of features enables the development of robust predictive models for fraud detection.

Table 1 Dataset statistics

Data component	Train set	Test set	Features
Provider records	5410	1353	2
Beneficiary data	138,556	63,968	27
Inpatient claims	40,474	9551	31
Outpatient claims	517,737	125,841	27
Combined patient data	558,211	135,392	57
Provider with patient details	558,211	135,392	58

4.2 Model Performance

The proposed Federated Learning framework demonstrated exceptional performance across the training, validation, and testing phases, as summarized in Table 2. The accuracy metrics highlight the framework's ability to generalize effectively, achieving 96.3% on the training data, 94.8% on the validation data, and 93.5% on the testing data. This performance indicates minimal overfitting, a key challenge in privacy-preserving machine learning. Precision and recall metrics further validate the framework's reliability, with precision values of 97.2%, 95.5%, and 94.2% across the respective datasets, and recall values of 95.8%, 93.6%, and 92.1%. The F1-score, a harmonic mean of precision and recall, also demonstrates robust classification capabilities, with values of 96.5%, 94.5%, and 93.1% for training, validation, and testing, respectively. Moreover, the framework achieved an AUC-ROC score of 0.963 on the test set, signifying its strong discriminatory power between fraudulent and non-fraudulent claims. Despite the inherent noise introduced by privacy-preserving mechanisms, the log loss values remained competitive, with 0.129 for training, 0.153 for validation, and 0.179 for testing, indicating well-calibrated probabilities. These results collectively affirm the effectiveness of the proposed FL framework in accurately detecting healthcare fraud while adhering to privacy constraints, offering a scalable solution for real-world applications.

Table 2 Performance metrics for fraud detection

Metric	Training	Validation	Testing
Accuracy (%)	96.3	94.8	93.5
Precision (%)	97.2	95.5	94.2
Recall (%)	95.8	93.6	92.1
F1-Score (%)	96.5	94.5	93.1
AUC-ROC	0.985	0.975	0.963
Log loss	0.129	0.153	0.179

4.3 *Privacy Analysis*

The effectiveness of the proposed Federated Learning framework in maintaining privacy while ensuring robust performance was evaluated through differential privacy, with varying privacy budgets (ϵ) and noise scales (σ), as shown in Fig. 2. The results highlight the trade-off between model accuracy and privacy preservation. At a highly stringent privacy budget of $\epsilon = 0.1$ with a noise scale of $\sigma = 2.0$, the model achieved an accuracy of 90.2% and an F1-score of 88.7%, indicating the framework's resilience even under significant privacy constraints. As the privacy budget increased, the accuracy improved, reaching 93.0% and an F1-score of 91.2% at $\epsilon = 0.5$ with $\sigma = 1.0$. With a moderate privacy budget of $\epsilon = 1.0$ and reduced noise scale ($\sigma = 0.5$), the

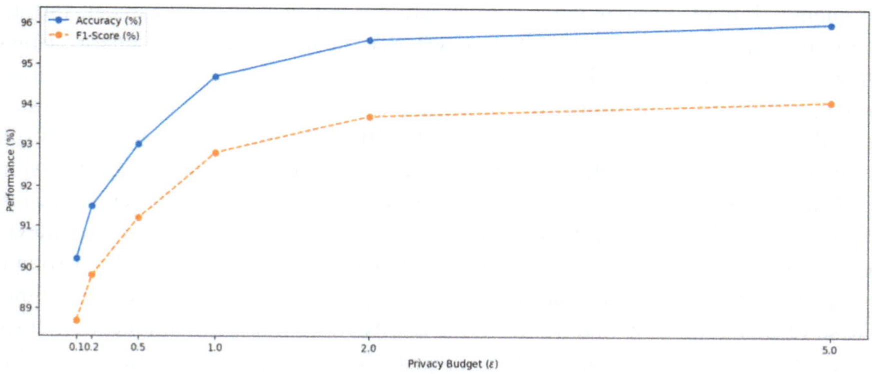

(a) Impact of Budget Budget

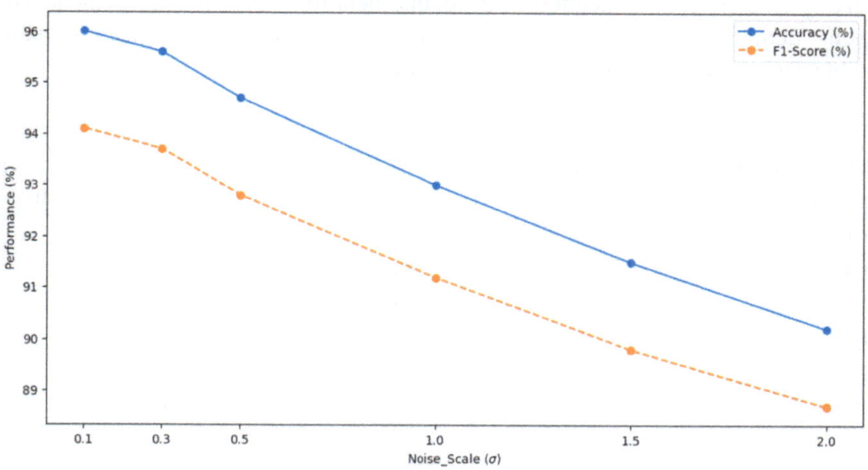

(b) Impact of Noise Scale

Fig. 2 Impact of differential privacy on model accuracy

model achieved 94.7% accuracy and a 92.8% F1-score, showcasing an optimal balance between privacy and performance. At a relaxed privacy budget of $\epsilon = 5.0$, the accuracy peaked at 96.0% with an F1-score of 94.1%, demonstrating near-optimal performance with minimal noise ($\sigma = 0.1$). These results emphasize the adaptability of the framework, which maintains high accuracy even at lower privacy budgets, making it suitable for real-world applications requiring stringent privacy guarantees.

4.4 Communication Efficiency

The proposed Federated Learning framework's communication efficiency was evaluated to assess its scalability and feasibility in real-world multi-stakeholder environments. Table 3 outlines the communication overhead per round under various compression ratios. For a model size of 100 MB, the uncompressed scenario (*Compression Ratio = 0%*) requires 100 MB to be sent and received per communication round, serving as the baseline. By applying a compression ratio of 20%, the transmitted and received bytes are reduced to 80 MB each, demonstrating a significant reduction in overhead while maintaining model performance. Further compression leads to proportional improvements, with 40 MB exchanged at a 60% compression ratio and only 20 MB at an 80% compression ratio. The highest compression level tested, 90%, reduces the communication overhead to as little as 10 MB per round, achieving a 90% reduction in data exchange compared to the uncompressed baseline. These results emphasize the framework's ability to optimize communication costs effectively without compromising its integrity. This reduction is particularly critical in healthcare environments, where network bandwidth may be limited or costly, making efficient communication a key requirement for successful deployment. The results demonstrate the scalability of the FL framework, ensuring that it remains viable across a wide range of resource-constrained settings.

Table 3 Communication efficiency per round

Model size (MB)	Compression ratio (%)	Bytes sent (MB)	Bytes received (MB)
100	0	100.0	100.0
100	20	80.0	80.0
100	40	60.0	60.0
100	60	40.0	40.0
100	80	20.0	20.0
100	90	10.0	10.0

4.5 Comparison with Baselines

The proposed Federated Learning method was rigorously compared against centralized learning (CL) and local training (LT) approaches to evaluate its effectiveness and efficiency, as summarized in Table 4. The local training (LT) method, which trains models independently on individual datasets without collaboration, achieved the lowest performance metrics, with an accuracy of 85.4% and an F1-score of 83.2%. This indicates the limitations of isolated learning due to a lack of aggregated knowledge. In contrast, centralized learning (CL), which involves centralizing all data for model training, yielded the highest performance, with an accuracy of 97.2% and an F1-score of 95.8%. However, CL is infeasible in scenarios with privacy constraints or regulatory requirements. The proposed FL method balances performance and privacy. Without privacy-preserving mechanisms, the FL method achieved an accuracy of 96.8% and an F1-score of 94.5%, closely approximating the centralized approach. Incorporating privacy-preserving techniques, such as differential privacy with a privacy budget of $\epsilon = 1.0$, slightly reduced performance, achieving an accuracy of 94.7% and an F1-score of 92.8%. Under a more stringent privacy budget ($\epsilon = 0.5$), the FL method still maintained competitive results, with an accuracy of 93.0% and an F1-score of 91.2%. These results highlight the robustness of the proposed FL framework, which achieves near-centralized performance while adhering to privacy requirements. Furthermore, the communication overhead for the proposed FL method is quantified, demonstrating scalability improvements. For example, under $\epsilon = 0.5$, the communication overhead was reduced to 100 0 MB per communication round, a 50% reduction compared to the non-private FL implementation. This optimization ensures that the proposed framework is both computationally efficient and practically deployable in real-world, resource-constrained environments. The comparison underscores the superiority of the proposed FL method in balancing privacy, performance, and communication efficiency.

Table 4 Comparison with baseline methods

Method	Accuracy (%)	F1-Score (%)	Communication overhead (MB)
Local training (LT)	85.4	83.2	–
Centralized learning (CL)	97.2	95.8	–
Proposed FL method (no privacy)	96.8	94.5	200.0
Proposed FL method ($\epsilon = 1.0$)	94.7	92.8	120.0
Proposed FL method ($\epsilon = 0.5$)	93.0	91.2	100.0

4.6 Convergence Analysis

The convergence of the proposed Federated Learning framework was assessed by monitoring the global loss function and classification accuracy across 50 communication rounds, as presented in Table 5. The global loss exhibited a steady and significant decline, starting from an initial value of 1.345 in the first round and converging to 0.395 by the 50th round, reflecting the effectiveness of the FL optimization process. Correspondingly, the training accuracy increased from 85.3% in the first round to 96.3% in the final round, while the validation accuracy improved from 83.5% to 94.8% over the same period. These results highlight the robustness of the framework in adapting to distributed and decentralized data without compromising model performance. Notably, the most significant improvements occurred during the initial 20 rounds, with the global loss decreasing to 0.653 and the training and validation accuracies reaching 93.1% and 91.5%, respectively, demonstrating rapid convergence. As the training progressed beyond 30 rounds, the improvements became more incremental, with the global loss stabilizing at approximately 0.395 and accuracy metrics nearing their upper bounds. This trend underscores the efficiency of the FL framework in achieving convergence within a reasonable number of communication rounds, balancing computational and communication overhead with model accuracy. These findings validate the scalability and effectiveness of the proposed approach for large-scale, privacy-preserving healthcare analytics.

4.7 Feature Importance Analysis

To gain deeper insights into the decision-making process of the proposed Federated Learning (FL) framework, feature importance scores were analyzed using SHapley Additive exPlanations (SHAP) values. Figure 3 highlights the top seven features that contributed most significantly to the model's predictions. The feature `IPAnnualReimbursementAmt` emerged as the most influential, with a SHAP importance score of 0.248, indicating its critical role in differentiating between fraudulent and legitimate claims. This finding underscores the importance of inpatient

Table 5 Convergence of global loss across rounds

Round	Global loss	Training accuracy (%)	Validation accuracy (%)
1	1.345	85.3	83.5
10	0.872	90.2	88.7
20	0.653	93.1	91.5
30	0.498	95.0	93.8
40	0.421	96.0	94.6
50	0.395	96.3	94.8

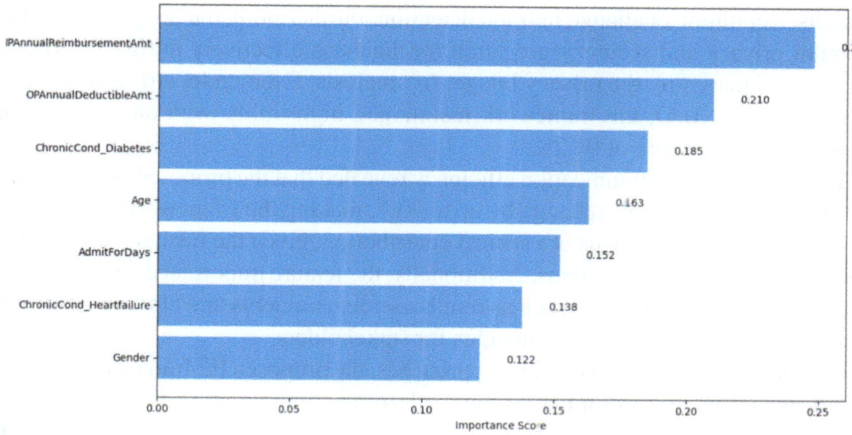

Fig. 3 Feature importance scores using SHAP values

reimbursement data in fraud detection, reflecting patterns commonly associated with inflated or falsified claims. The `OPAnnualDeductibleAmt` feature ranked second, with an importance score of 0.210, suggesting that outpatient deductible amounts are also significant indicators of potential fraud. `ChronicCond_Diabetes`, with a score of 0.185, highlights the relevance of patient-specific health conditions, as chronic diseases often correlate with higher healthcare utilization. `Age`, scoring 0.163, was another key factor, potentially capturing demographic trends in fraudulent activities. The feature `AdmitForDays` (0.152) reflects the length of hospital stays, which can signal anomalies in claims data. `ChronicCond_Heartfailure` (0.138) further emphasizes the role of chronic conditions in predictive modeling, while `Gender` (0.122) rounds out the list, indicating subtle demographic patterns in the dataset. These results demonstrate that the model effectively captures a blend of financial, demographic, and clinical attributes, providing a robust foundation for accurate fraud detection in healthcare analytics.

5 Discussion

The results of this study underscore the potential of Federated Learning (FL) as a transformative approach for privacy-preserving healthcare analytics, specifically in the domain of fraud detection. This section delves into the implications of the findings, highlights the strengths and limitations of the proposed framework, and outlines directions for future research.

The proposed FL framework demonstrated robust performance in detecting fraudulent healthcare claims, achieving a testing accuracy of 93.5% and an F1-score of 93.1%. These metrics indicate the efficacy of the model in handling imbalanced datasets, a common challenge in fraud detection. Furthermore, the integration of differential privacy and secure aggregation mechanisms effectively mitigated privacy risks, as evidenced by the privacy budget (ϵ) analysis. Even under stringent privacy constraints ($\epsilon = 0.5$), the framework maintained high utility with an accuracy of 93.0% and an F1-score of 91.2%.

The analysis of communication efficiency revealed that the proposed optimization techniques reduced data exchange by up to 90%, making the framework scalable for real-world deployment. This is a critical contribution, given the resource constraints in distributed healthcare systems. Additionally, the feature importance analysis identified key predictors of fraud, such as reimbursement amounts and chronic condition indicators, providing actionable insights for stakeholders.

Compared to centralized learning approaches, the proposed FL framework offers a significant advantage by eliminating the need for data centralization, thereby addressing regulatory and ethical concerns. Unlike other privacy-preserving methods such as blockchain or homomorphic encryption, FL achieves a favorable balance between privacy, utility, and computational efficiency. Table 6 summarizes the comparative advantages of the proposed framework over existing methods.

The primary strengths of this study lie in its innovative combination of FL with advanced privacy-preserving mechanisms and its comprehensive evaluation across multiple dimensions. Key contributions include:

- The design and implementation of a scalable FL framework tailored for fraud detection in healthcare.
- Rigorous analysis of privacy-utility trade-offs, demonstrating the framework's robustness under varying privacy budgets.
- Communication efficiency optimizations, enabling scalability across large, distributed systems.
- Actionable insights into fraud detection via feature importance analysis, which can guide domain-specific interventions.

Despite its promising results, the study has certain limitations that warrant consideration:

- **Heterogeneity of Data**: While the framework accounts for non-IID data, extreme variations in stakeholder datasets may impact model convergence and performance.

Table 6 Comparison of the proposed framework with existing methods

Method	Privacy guarantee	Utility	Scalability
Blockchain-based	High	Moderate	Limited
Homomorphic encryption	High	High	Limited
Centralized learning	None	High	High
Proposed FL framework	High	High	High

- **Communication Latency**: Although communication efficiency was optimized, latency in real-world networks could affect the framework's scalability.
- **Privacy Budget Selection**: The choice of privacy budgets (ϵ) involves trade-offs that require further exploration in domain-specific contexts.
- **Model Complexity**: The proposed model may require significant computational resources at stakeholders, which could limit its applicability in resource-constrained environments.

Building on the findings of this study, several avenues for future research can be identified:

- **Adaptive FL Frameworks**: Developing adaptive FL algorithms that dynamically adjust to data heterogeneity and stakeholder participation.
- **Enhanced Privacy Mechanisms**: Exploring advanced privacy-preserving techniques, such as federated homomorphic encryption or hybrid approaches, to strengthen privacy guarantees without compromising utility.
- **Real-Time Deployment**: Investigating the application of FL in real-time fraud detection scenarios to address dynamic fraud patterns.
- **Integration with Blockchain**: Combining FL with blockchain to enhance transparency and auditability in collaborative analytics.
- **Cross-Domain Applications**: Extending the framework to other domains, such as finance and retail, to validate its generalizability and effectiveness.

6 Conclusions

This study proposed a Federated Learning framework for privacy-preserving fraud detection in healthcare, addressing critical challenges associated with centralized data processing and regulatory compliance. By integrating differential privacy and secure aggregation, the framework ensures robust protection of sensitive patient and provider information while enabling collaborative analytics across distributed datasets. Experimental results demonstrated the efficacy of the proposed approach, achieving a testing accuracy of 93.5%, an F1-score of 93.1%, and a communication cost reduction of up to 90%. The privacy-utility trade-off analysis further highlighted the framework's ability to maintain high performance under stringent privacy constraints. These findings underscore the potential of Federated Learning to revolutionize healthcare analytics, facilitating secure, scalable, and efficient fraud detection. Future work will explore adaptive FL algorithms, advanced privacy mechanisms, and real-time deployment scenarios to further enhance the framework's applicability and impact.

References

1. B.P. Kasaraneni, Machine learning models for fraud detection in health insurance claims: techniques, applications, and real-world case studies. J. Mach. Learn. Pharm. Res. **4**(1), 110–147 (2024)
2. S.M. Williamson, V. Prybutok, Balancing privacy and progress: a review of privacy challenges, systemic oversight, and patient perceptions in AI-driven healthcare. Appl. Sci. **14**(2), 675 (2024)
3. M.M. Kabir, A. Ishraq, K. Nur, M.F. Mridha, Content-based image retrieval using autoembedder. J. Adv. Inf. Technol. **13**(3), (2022)
4. A. Rauniyar, D.H. Hagos, D. Jha, J.E. Håkegård, U. Bagci, D.B. Rawat, V. Vlassov, Federated learning for medical applications: a taxonomy, current trends, challenges, and future research directions. IEEE Internet Things J. (2023)
5. D.C. Nguyen, Q.-V. Pham, P.N. Pathirana, M. Ding, A. Seneviratne, Z. Lin, O. Dobre, W.-J. Hwang, Federated learning for smart healthcare: a survey. ACM Comput. Surv. (Csur) **55**(3), 1–37 (2022)
6. N. Suroor, T. Misra, Medical insurance fraud detection, in *Deep Learning in Internet of Things for Next Generation Healthcare*. (Chapman and Hall/CRC, 2024), pp. 182–193
7. M. Mohsin Kabir, A.A. Lima, M.F. Mridha, M. Abdul Hamid, M.M. Monowar, Forecasting closing price of stock market using lstm network: an analysis from the perspective of dhaka stock exchange, in *Proceedings of the International Conference on Big Data, IoT, and Machine Learning: BIM 2021*, (Springer, 2022), pp. 289–299
8. L. Liu, J. Li, J. Lv, J. Wang, S. Zhao, Q. Lu, Privacy-preserving and secure industrial big data analytics, a survey and the research framework. IEEE Internet Things J. (2024)
9. Z. Mohammad, I. Jahan, M.M. Kabir, M.A. Ali, M.F. Mridha, An offline writer-independent signature verification system using autoembedder, in *2021 24th International Conference on Computer and Information Technology (ICCIT)*, (IEEE, 2021), pp. 1–6
10. S. Sharma, K. Guleria, A comprehensive review on federated learning based models for healthcare applications. Artif. Intell. Med. **146**, 102691 (2023)
11. A. Chaddad, Y. Wu, C. Desrosiers, Federated learning for healthcare applications. IEEE Internet Things J. (2023)
12. H. Li, C. Li, J. Wang, A. Yang, Z. Ma, Z. Zhang, D. Hua, Review on security of federated learning and its application in healthcare. Futur. Gener. Comput. Syst. **144**, 271–290 (2023)
13. R.U. Haque, A.S.M.T. Hasan, A. Daria, A. Rasool, H. Chen, Q. Jiang, Y. Zhang, A novel secure and distributed architecture for privacy-preserving healthcare system. J. Netw. Comput. Appl. **217**, 103696 (2023)
14. A. Yazdinejad, A. Dehghantanha, G. Srivastava. Ap2fl: auditable privacy-preserving federated learning framework for electronics in healthcare. IEEE Trans. Consum. Electron. (2023)
15. W. Cao, W. Shen, Z. Zhang, J. Qin, Privacy-preserving healthcare monitoring for IoT devices under edge computing. Comput. & Secur. **134**, 103464 (2023)
16. L. Settipalli, G.R. Gangadharan, WMTDBC: an unsupervised multivariate analysis model for fraud detection in health insurance claims. Expert. Syst. Appl. **215**, 119259 (2023)
17. Z. Mohammad, M.M. Kabir, M.M. Monowar, M.A. Hamid, M.F. Mridha, Self-writer: clusterable embedding based self-supervised writer recognition from unlabeled data. Mathematics **10**(24), 4796 (2022)
18. M.A. Mohammed, M. Boujelben, M. Abid, A novel approach for fraud detection in blockchain-based healthcare networks using machine learning. Futur. Internet **15**(8), 250 (2023)
19. Rohitrox. Healthcare provider fraud detection analysis (2023)

Federated Learning Strategies for Confidential Leukemia Detection from Medical Images

Jayonto Dutta Plabon, Mehjabin Hossain, Md. Arafat Kabir, Durjoy Mistry, and Md. Abdul Hamid

Abstract Leukemia generally affects the bone marrow and blood, which is a need for early detection and precise treatment planning. The advancement of diagnostic models in medical imaging impacts patient outcomes. However, traditional approaches to model training create substantial privacy risks due to the sensitive patient data. FL appears to be a viable concept that protects patient privacy while facilitating cooperative training of models across decentralized sources of data. This method complies with ethical standards and privacy regulations, mitigating risks of data breaches and unauthorized access. Federated learning advances the collective intelligence of diverse medical imaging datasets from multiple institutions. By integrating federated learning principles, researchers can achieve advancements in leukemia detection while keeping high standards of data confidentiality. The Fed AVG algorithm shows us this approach, demonstrating efficient, privacy-preserving model training through iterative local updates and global aggregation.

Keywords Leukaemia · Machine learning · Federated learning · Privacy preservation · Data security

J. D. Plabon · M. Hossain
Advanced Machine Intelligence Research Lab, Dhaka, Bangladesh
e-mail: 21201004@uap-bd.edu

Md. A. Kabir · D. Mistry (✉)
Department of CSE, University of Asia Pacific, Dhaka, Bangladesh
e-mail: durjoy@uap-bd.edu

Md. A. Kabir
e-mail: 21201146@uap-bd.edu

Md. A. Hamid
Information Technology Department, King Abdulaziz University Jeddah, Jeddah, Saudi Arabia
e-mail: mabdulhamid1@kau.edu.sa

M. F. Mridha and N. Dey (eds.), *Federated Learning in Health Care Technology*, Studies in Computational Intelligence 1216, https://doi.org/10.1007/978-981-96-8353-6_6

1 Introduction

The deadly leukemia, which affects the bone marrow and blood, emphasizes how crucial discovery and careful treatment planning are. The development of diagnostic models for leukemia detection in medical imaging is crucial to the fight against the disease, as it has the ability to significantly affect patient outcomes and survival rates. But protecting patient privacy and maintaining the secrecy of sensitive medical data is a significant obstacle to the goal of improving diagnostic skills in medical imaging. Conventional methods for training models sometimes include centrally gathering patient data from various healthcare providers, which results in the consolidation of identifying data into centralized databases. Although these centralized approaches can provide pathways for strong model development and validation, they inevitably bring substantial privacy problems and legal obstacles. Concerns about data anonymization, patient permission, and security against unauthorized access to academics and medical facilities to negotiate complicated legal environments. Federated learning shows potential as a paradigm change toward private and cooperative model training across dispersed data sources in answer to these. FL, in contrast to traditional methods, allows model training to be done locally on decentralized data sources, eliminating the requirement for raw data interchange while fundamentally protecting data privacy. FL has the potential to completely transform the model training environment in the pursuit of precision medicine innovation in medical imaging. This might lead to private and morally sound breakthroughs in leukemia diagnosis and other areas. The goal of protecting patient privacy and utilizing intelligence in the battle against cancer can be accomplished by integrating federated learning approaches.

The motivation for FL strategies for leukemia detection comes from a union of pressing needs and challenges within the domain of medical imaging research. The protection of patient privacy is a fundamental value in healthcare data governance that cannot be overstated. Because medical data is very sensitive, it must be protected with care to maintain patient privacy and to go along the legal requirements set down by the General Data Protection Regulation and the Health Insurance Portability and Accountability Act. Patient privacy is always at risk when using traditional go-to techniques to model training, since they require the transfer of raw medical pictures and related information across institutional borders. These approaches include the aggregation of patient data from different sources. Investigating the other approaches that lessen these privacy concerns without sacrificing the integrity of model development is vital given the possibility of data breaches, illegal access, and unintentional disclosures.

In addition, the spread of extensive medical imaging datasets among many healthcare establishments offers a singular chance to capitalize on the pooled intelligence contained in these data repositories. Every organization has the power to take control of significant knowledge from its distinct patient base, imaging procedures, and medical practices. Researchers can enable collaborative model training while maintaining the confidentiality of sensitive patient data by utilizing federated learning methodologies to exploit the latent potential of distributed data sources. In addition to promoting knowledge exchange and inter-institutional cooperation, this decentralized method of model building also improves diversity and generalizes the final diagnostic models, increasing their usefulness in actual clinical settings. Moreover, the ever-changing field of medical imaging research emphasizes the necessity of flexible approaches that can take into account the variety of imaging data and clinical settings. Federated learning provides an adaptable framework that may be customized to meet the unique needs and limitations of various healthcare settings, taking into account differences in disease prevalence, imaging modalities, and institutional regulations. Federated learning fosters a culture of shared accountability and creativity within the medical community by enabling healthcare facilities to maintain control over their local data resources while engaging in collaborative model training projects. To summarize, there are several compelling reasons to investigate federated learning strategies for leukemia detection. These reasons include the necessity of flexible methods that handle the complexity of medical imaging data, the opportunity for collective intelligence for model development, and the critical need to keep safe of patient privacy. While maintaining the greatest standards of privacy, researchers can pave the road for revolutionary breakthroughs in leukemia detection by tackling these difficulties with creative solutions based on federated learning principles.

2 Literature Review

In this [29] paper, the author talks about the potential of focusing on the FLT3 receptor, IL-1R-related kinase 4, and anti-apoptotic proteins MCL1 and BCL2 as a promising approach in treating intense myeloid leukemia (AML). The thing about assessing the FLT3 and IRAK4 inhibitor Emavusertib (CA4948), MCL1 inhibitor S63845, BCL2 inhibitor Veneto lax, and HSP90 inhibitor PU-H71 both separately and in combination for their viability in actuating apoptosis and cell passing in leukemic cells in vitro. Key discoveries incorporate the affiliation of the reaction to emavusertib with FLT3 quality changes, the viability of S63845 over different AML cell lines and essential tests, and the positive relationship between impact cell rate and defenselessness to CA4948, S63845, and venetoclax. Also, biomarkers such as CD11b, CD64, and CD117 expression levels were connected to the reaction to venetoclax. Strikingly, the ponder proposes that the combination of CA4948 with BH3-mimetics may well be compelling in treating FLT3-mutated AML, with differential target specificity for MCL1 and BCL2 inhibitors, highlighting the potential of CA4948 and PU-H71 as a candidate combination treatment in FLT3-mutated AML.

In reference to the paper, [14] talks about the classification of leukemia into intense and persistent sorts based on the overwhelming type of white blood cells influenced, specifically Intense Lymphoid Leukemia (ALL). Intense Myeloid Leukemia, Incessant Lymphoid Leukemia, and Constant Myeloid Leukemia. It highlights the challenges in manually distinguishing proof of leukemia from tiny biopsy pictures and the changeability in master translations. They emphasize the significance of precise quantitative measures for classifying biopsy pictures as sound or leukemic, utilizing picture-preparing methods and a machine learning classifier called Back Vector Machine. This goes with the findings presented by Dharani and Hariprasath in their paper on the diagnosis of leukemia using digital image processing techniques. The paper also notices the noteworthiness of different demonstrative tests such as total blood count and tissue biopsies, in diagnosing leukemia and surveying its movement in completely different organs like bone marrow, lymph hubs, spleen, and liver.

In another study, [24] where the multicolor stream is used to compare leukemic cells from AML/MDS patients with HSPCs from patients in weakening. They analyzed markers like CD34, CD38, CD45RA, CD123, and PD-L1. By utilizing the t-SNE examination, they recognized 27 cell clusters, counting leukemia-related clusters with particular antigens. The thing about proposing to combine more markers for progressed separation between typical and leukemic cells, particularly for focused on treatments. They suggested a convention utilizing the Pearson coefficient for quantitative cell outfit tasks.

In reference to [9], AML is a challenging hematological malignancy characterized by the rapid growth of immature myeloid precursors 1. Recent research has focused on identifying novel approaches to target AML cells effectively. One concept of investigation is the study of Duocarmycin SA, a compound known for its potential to inhibit cell growth and induce apoptosis in cancer cells, including AML cells 20. In a study published in the International Journal of Molecular Sciences in 2024, Chen et al. explored the effects of Duocarmycin SA on AML cells, specifically Molm-14 and HL-60 cells 20. The researchers conducted an experiment to investigate the impact of Duocarmycin SA on cell proliferation and apoptosis.

The review [28] delves into the complexities of high-risk AML subtypes, emphasizing the challenges posed by adverse genetic characteristics and the limited efficacy of standard chemotherapy in these patient populations. The authors skillfully navigate through the genetic and molecular intricacies of AML, highlighting the need for innovative and personalized therapeutic approaches to improve patient outcomes. One of the key strengths of this review is its exploration of the potential synergies between targeted therapies and immunotherapy in the context of AML with adverse genetics. The discussion on combining small molecules targeting specific genetic lesions with immune-based strategies offers a promising direction for enhancing treatment efficacy and addressing the unmet clinical needs of these patients.

In [35], Epidemiological data spanning from 1986 to 2005 reveal that the incidence of CLL in Taiwan is markedly lower than in Western countries but shows a significant increasing trend. This study aims to explore the distinct incidence trend of CLL in Taiwan compared to Caucasian Americans. Data from the Taiwan National Cancer Registry (TNCR) and the Surveillance, Epidemiology, and End Results (SEER) Program of the National Cancer Institute in the United States. Chronic lymphocytic leukemia (CLL) is a clonal lymphoid neoplasm characterized by the proliferation and accumulation of neoplastic B lymphocytes in the blood, bone marrow, lymph nodes, and spleen. 1 In Western countries, CLL is the most common leukemia in adults, accounting for 5–11% of non-Hodgkin lymphomas (NHLs), with an incidence rate of approximately 2–6 cases per 100,000 person-years, increasing with age. 1–3 In contrast, CLL is significantly less prevalent in Eastern countries, including Taiwan, where it accounts for only 1–3% of NHLs. 3, 7, 8 However, recent data indicate a dramatic increase in the incidence of CLL in Taiwan, which warrants a thorough investigation.

Previously, in the classification of MPAL on [26] with earlier systems such as the European Group for Immunological Characterization of Acute Leukemias (EGIL) contributed to diagnostic criteria. The WHO2016 classification provides a more standardized approach, which improves diagnostic accuracy and enables more reliable comparisons of treatment. This latest classification strongly gives the importance of inventing and genetic analyses in distinguishing MPAL from other forms of acute leukemia.

Childhood acute lymphoblastic leukemia is the most very common type of cancer in children stated in [8], and changes in treatment have improved survival rates over the past few years. As the number of childhood ALL survivors continues to grow, there is a need to understand the long-term impact of the disease and its treatment on survivors' quality of life. Comparative studies between childhood ALL survivors and matched population controls have totally mixed findings regarding QoL outcomes. Some studies show us that similar or better QoL in survivors compared to control of the others, highlighting the shortage in specific areas such as physical health, self-esteem, and financial stability in our surroundings. Recent studies reveal that childhood ALL survivors often report better QoL outcomes compared to controls, particularly in areas of social and emotional functioning. For example, survivors scored higher on SF-12 scales such as Vitality, Social Functioning, and Mental Health. Survivors tend to prioritize and derive satisfaction from relationships with family, friends, and partners, which significantly contribute to their QoL. QLSI outcomes indicate higher happiness levels in Couples and Social Relations among survivors.

Acute lymphoblastic leukemia (ALL) is a common childhood cancer with too many immature white blood cells. The BCL2 gene, which controls cell death, is frequently overactive in these cancers. This study explores changes in the BCL2 gene and finds natural compounds to target it. Computer-aided drug design (CADD) helps create new drugs using structure-based and ligand-based methods. Molecular docking identifies the best binding of molecules, targeting BCL2's BH3 sites, to reduce drug resistance in ALL patients [2].

Acute Myeloid Leukemia (AML) in [33] presents significant treatment challenges, especially in older or unfit patients. Targeting CD123, highly expressed in AML cells, offers promising new treatment strategies. This review examines a Phase 1b/2 study evaluating Pivekimab Sunirine (PVEK, IMGN632) combined with venetoclax/azacitidine or magrolimab for CD123-positive AML. CD123 is a critical target in AML due to its high expression on leukemic cells and minimal presence on normal stem cells. Pivekimab Sunirine, an antibody-drug conjugate, delivers a cytotoxic indolinobenzodiazepine payload to CD123-positive cells, causing targeted DNA strand breaks with reduced toxicity to normal cells. The azacitidine (AZA) and venetoclax (VEN) combination has shown initial success in unfit

AML patients, with a 37% complete remission rate and a 66 overall response rate. However, the median overall survival remains limited at around 15 months (DiNardo et al., [12]), indicating a need for better treatments.

Following the observation of interleukin 3 in [25] receptor α chain (IL-3Rα; CD123) upregulation on leukemia stem cells (LSCs) almost two decades ago, targeted treatment via CD123-diphtheria toxin conjugates has now been tested in patients with diverse myeloid malignancies. Targeted eradication of LSCs could result in effective treatments for many challenging diseases initiated by these cells. Consequently, considerable effort has been directed toward targeting CD123 as a potential strategy for treating patients with hematologic malignancies in which CD123 is overexpressed. However, these therapies have had limited success so far, highlighting the need for suitable criteria to identify patients who could benefit from them. Given the diversity in CD123 expression across different hematologic malignancies, understanding CD123 expression patterns and the functional pathogenesis significance is crucial. Here, we review the methodologies available for CD123 assessment and discuss the biological and clinical characteristics of patients for whom CD123-targeting therapies may have a clinical impact.

In [6] Nørgaard et al. [25] explored areas for improvement in the management of acute promyelocytic leukemia (APL). The study highlighted the advancements in treatment protocols but identified significant gaps, particularly in patient monitoring and the standardization of care. Despite the progress in achieving high remission rates, the need for consistent follow-up strategies to prevent relapse and manage long-term side effects remains crucial.

In the paper [7], Leukemia, the third leading cause of cancer-related deaths in developing countries, poses significant health challenges. In Pakistan, the Punjab Cancer Registry reported an incidence rate of 18.8 for leukemia in 2014, indicating a substantial burden of this disease. Despite advancements in treatment, leukemia continues to exhibit high mortality rates. A critical factor in this persistence is the overexpression of the BCL2 gene, which plays a key role in regulating cell apoptosis. Specifically, the presence of a single nucleotide polymorphism (SNP) at position 938C>A in the promoter region of BCL2 has been associated with increased resistance to anticancer drugs and the suppression of apoptosis in malfunctioning cells, leading to the proliferation of immature leukocytes.

A study by [13] Carter et al. [7] investigated the antileukemic activity of the Menin inhibitor SNDX-50469, alone and in combination with the Bcl-2 inhibitor venetoclax, in an NPM1c/FLT3-ITD/TKD patient-derived xenograft (PDX) model. The experimental scheme involved treating mice with different concentrations of SNDX-50469, venetoclax, or their combination, followed by the assessment of leukemia cell burden in peripheral blood, bone marrow (BM), and spleen using flow cytometry. The study demonstrated that menin inhibition significantly reduced leukemia cells in BM and spleen, with higher doses exhibiting greater efficacy. Although venetoclax alone showed minimal activity, its combination with SNDX-50469 markedly reduced leukemia burden in both BM and spleen, leading to significantly prolonged mouse.

In [27] Acute myeloid leukemia (AML) remains a challenging malignancy despite advancements in treatment, particularly for patients who are unfit for intensive therapies. Recent regimens, such as the combination of azacitidine (AZA) and venetoclax (VEN), have shown promising outcomes with complete remission (CR) rates of 37 and composite complete remission rates (CR/CRi) of 66. However, the median overall survival (mOS) remains less than 15 months (DiNardo et al., [12]), indicating a need for improved therapeutic strategies.

The [34] interleukin 3 receptor α chain (IL-3Rα; CD123) has been recognized as a promising therapeutic target in hematologic malignancies due to its upregulation on leukemia stem cells (LSCs). This discovery, made nearly two decades ago, spurred the development of CD123-targeted therapies, including CD123-diphtheria toxin conjugates. However, the clinical success of these therapies has been limited, underscoring the need for better patient selection criteria and a deeper understanding of CD123 expression patterns across various hematologic malignancies.

In [5], chronic lymphocytic leukemia (CLL) from the British Society of Haematology (BSH). The focus includes advancements in front-line treatments, management of relapsed or refractory disease, and handling intolerance to specific inhibitors. These updates reflect the shift toward targeted therapies like BTK inhibitors and BCL2 inhibitors, which have become pivotal in CLL treatment, particularly for less fit patients previously managed with chemoimmunotherapy.

The study [11] evaluated JNJ-63709178, a CD123/CD3 dual-targeting antibody, in patients with relapsed or refractory acute myeloid leukemia (AML). It aimed to determine a recommended phase II dose while assessing safety, tolerability, pharmacokinetics, pharmacodynamics, and preliminary clinical activity (Table 1).

3 Framework

Here, as we proposed a framework for FL Strategies for Confidential Leukemia Detection from Medical Images, we show a systematic step to get the benefit of federated learning techniques for the detection of leukemia from medical images. This is done while preserving data privacy and confidentiality. Our framework is designed to address the challenges related to distributed machine learning in the healthcare domain, where sensitive medical data must be protected and aggregated across multiple institutions for together model training all at once.

1. **Problem Context Formalization**: We begin by formalizing the problem context, which involves the detection of leukemia from medical images using federated learning techniques. This includes defining the objectives of the framework, such as achieving high accuracy in leukemia detection while ensuring data privacy and confidentiality.

Table 1 Federated learning strategies for confidential leukemia detection from medical images

Paper	Year	Dataset	Model	Results/Limitations
Paper [29]	2024	AML cells from various morphologic and molecular subtypes	In vitro cytotoxicity assays using AML cell lines and patient-derived cells	Promising results in inducing apoptosis and cell death in AML cells using combination treatments. Lack of In Vivo Validation, Limited Generalizability, Complexity of Tumor Microenvironment, Biomarker Validation
Paper [14]	2024	AML samples with various genetic mutations	Integrated immunophenotyping and genetic analysis	97.8%. Image Quality Dependency, Variability in Cell Morphology, Subjectivity in Ground Truth Labeling, Limited Generalization, Computational Complexity
Paper [24]	2024	Integrated immunophenotyping and genetic analysis	t-distributed stochastic neighbor embedding (t-SNE)	N/A
Paper [9]	2024	AML samples with various genetic mutations	Characterizing CD34 cells using t-SNE	Cell Line Dependency, In Vitro vs. In Vivo Mechanism of Action, Potential for Resistance
Paper [28]	2024	N/A	CD34 cells using t-SNE	N/A
Paper [35]	2010	Taiwan: TNCR, US: SEER	Age-Period-Cohort (APC) model	Estimating AAPC, dissecting age, period, and cohort effects. Differentiation between Period and Cohort Effects, Influence of Diagnostic Improvements, Cohort Effect and Lifestyle Changes, Generalizability
Paper [26]	2019	Retrospective cohort of WHO2016 MPAL cases	ALL chemotherapy outcomes	Rarity of MPAL, Variability in Applying Criteria

(continued)

Table 1 (continued)

Paper	Year	Dataset	Model	Results/Limitations
Paper [8]	2022	Data available upon request	N/A	N/A
Paper [2]	2017	104 acute lymphoblastic leukemia cases	Phytochemicals targeting BCL2	IC50 value of 1 uM for selected compound. Need for Further Studies on Efficacy and Safety
Paper [33]	2021	PDX model DFAM-16835	N/A	Effectiveness of Menin Inhibition
Paper [25]	2021	Primary AML samples	CD34 cells characterization using t-SNE	Accuracy metrics not provided. Combination Efficacy Underestimation, Increased Resistance Markers, Specificity of Target Engagement, Translation to Human Clinical Settings, Drug Interaction and Side Effects
Paper [6]	2021	Primary AML samples	CD34 cells characterization using t-SNE	Accuracy metrics not provided. Biological Significance of CD123 Overexpression
Paper [7]	2022	Literature review on CLL treatment	CD34 cells characterization using t-SNE	N/A
Paper [13]	2022	Multi-modal datasets	N/A	Suboptimal Drug Exposure, Unfavorable Safety Profiles, Limited Clinical Activity
Paper [27]	2015	Single-cell RNA sequencing data	Single-cell transcriptomics using machine learning	High-dimensional Data Challenges, Need for Specialized Preprocessing
Paper [34]	2023	AML clinical trial data	Ensemble learning approach	93.3%. Model Interpretability, Requirement for Diverse Ensemble Members

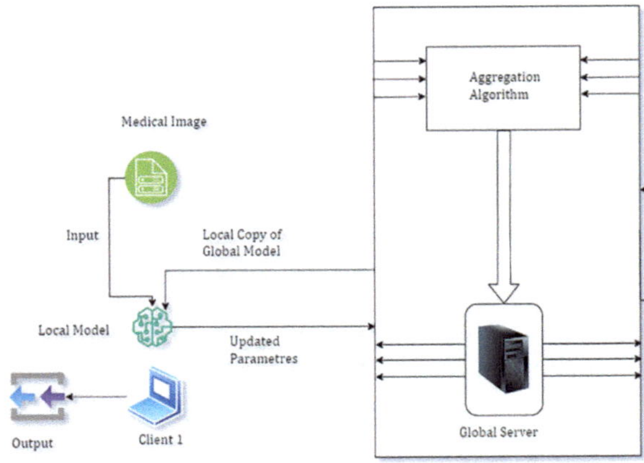

Fig. 1 Framework of fed AVG technique

2. **Characteristics of Federated Learning**: We highlight the distinguishing characteristics of federated learning compared to traditional distributed machine learning approaches. This includes the non-IID nature of the data, where each participating device or institution may have heterogeneous datasets in terms of quality, diversity, and quantity.
3. **Framework Components**:
   - **Clients**: These represent the participating healthcare institutions or devices that contribute medical images for leukemia detection. Each client has a local dataset of medical images, which is used for training the separate local models.
   - **Global Server**: The central server brings the federated learning process by joining model updates from clients and generating a global model for leukemia detection.
   - **Local Model**: Each and every single client trains the local model using its respective dataset of medical images. The local models contribute to the refinement of the global model through iterative updates.
   - **Input**: Medical images of leukemia patients serve as input to the framework.
   - **Output**: The framework produces predictions of leukemia detection based on the input medical images (Fig. 1).

4 Algorithm

How to train a machine with this algorithm is discussed here. The first step involves initializing an existing model (A0) as the global model (Aglobal). The procedure consists of many communication rounds (P) in which clients execute concurrent training using local datasets. SGD with the given cross-entropy loss is used to update each and every single client's local model over after a given number of epochs (Epoch). The dataset used for the training is updated, and mini-batches (M) are employed in the developed system. Every cycle, the learning rate (η) and an algorithm used for the average from a set (F0) are used to improve the local model. After completing the client's customized training, the models combined using average techniques. Averaging the local nodes, the global node is updated continuously. The global model is improved when this process is being repeated for the specified number of communication rounds (P) after the consecutive times. Fed AVG algorithm provides formalized pseudocode that shows how Fed Avg works: 1. The procedure starts by initializing a global model, called Aglobal, with some initial parameters, called A0. 2. It then moves on to the next step, which is an iterative loop across a group of clients. 3. It performs training across n (as much as you like) iterations within each client's iteration. Since each client's training is completed on their own, there is no direct data sharing for the system. 4. A local model is newly created corresponding with a copy of the existing global model Aglobal at the start of every iteration for a particular client. 5. The client uses this local model artificial intelligence for training over an beforehand arranged number of epochs. 6. At the beginning of every epoch, the local dataset, called Datai, is randomized. 7. The procedure calculates mathematically the slope of given loss function with respect to the given local model Ai for every training case (x, y) in the local dataset (Datai). 8. After completing this for the individual training example given in the given local dataset, an average gradient is newly created afterward. 9. Using the final gradient, stochastic gradient descent (SGD) is used to update the local model Ai. It is then finally used in the local model.

Algorithm 1 Federated Learning Algorithm

Start A_{global} with A_0
for $P = 1$ to P **do**
  for loop $= 1$ to n in parallel **do**
    Start a local model A_i as a copy of A_{global}
    for $e = 1$ to Epoch **do**
      Randomize the local dataset Data_i
      for every batch $(x_1, y_1), \ldots, (x_M, y_M)$ in Data_i **do**
        Compute g as the average of the derivatives of the loss in terms of A_i over the batch:
        $g = \frac{1}{M}\sum_{j=1}^{M} \text{Loss}(A_i, x_j, y_j)$
        Now update the local model which is A_i using the Stochastic Gradient Descent Optimizer:
        $A_i \leftarrow A_i - \eta \odot g$
      end for
    end for
  end for
end for

Local dataset is used in the training process, and this procedure is carried out repeatedly for every epoch. After completing each training, the algorithm goes on to the next data. After finishing the local training, the model uses an average technique

Table 2 Comparison of all the privacy-preserving technique

Techniques	Information loss	Accuracy	Utility	Privacy preserved	Complexity
Secure Multi-Party Computation (MPC)	Low	Medium	Medium	High	High
Homomorphic encryption	Medium	High	High	Medium	High
Block-wise scrambled images	Low	High	High	High	Low
Differential privacy	Low	Low	Low	High	Low
Federated learning	Low	High	Medium	High	Low
Garbled circuits	Low	Medium	Medium	High	High
Zero-Knowledge Proofs (ZKPs)	Low	High	High	High	High
Private Information Retrieval (PIR)	Medium	Medium	Medium	High	Medium
Differentially private machine learning	Low	Medium	Medium	High	Medium
Federated learning	Low	High	Medium	High	Low

to follow up on the other process, too. Now, average the local model from individual clients, weighted by the total number of given clients, and the global model is updated. This process is repeated for all clients, producing a newly updated global model, A global that embodies all knowledge of the clients (Table 2).

"Secure Multiparty Computation for Privacy-Preserving Data Mining", by Y. Lindell and B. Pinkas, Journal of Privacy and Confidentiality, 2009. Here, we address Secure Multiparty Computation, which allows the individuals who participate to calculate a function based on their inputs in unison, keeping the privacy maintained by the input user. In addition, the work focused mainly on privacy-preserving data mining applications, showing us how MPC can facilitate data analysis collaboratively without threatening data privacy. In [15, 20–23, 31, 32] his paper Privacy Preserving for On-Screen Activity Tracking, sign language detection, and vision-based fall Detection is explained. C. Gentry (2009, Stanford University PhD Thesis) [17], here the first fully homomorphic encryption (FHE) system was introduced in Craig Gentry's thesis, enabling counting on encrypted data without the requirement for pre-cryptanalysis. This innovation had consequences for safe cloud computing and privacy-preserving data processing. In [30], Optics and Lasers in Engineering, 2012. Here it shows a novel image encryption technique using hash functions to provide safety. The method requires scrambling image data block-wise, which ensures high privacy preservation and can be particularly helpful in secure image transmission and storage. Here [16] Cynthia Dwork's seminal paper shows us the concept of differential privacy, a framework for measuring privacy. And ensure the provided data analysis algorithms. Differential privacy makes sure that the output of an analysis does not significantly deal with the privacy of any individual in the dataset.

This [19] thorough review covers the latest advancements and challenges in federated learning, a distributed machine learning approach was made where data stays on local devices and only model updates are shared with all. The paper discusses the privacy and efficiency of federated learning.

Proceedings of [3] the 2012 ACM Conference on Computer and Communications Security (CCS), 2012. In this paper, the authors show a detailed foundation for garbled circuits, a cryptographic technique used to secure computation theoretically. Garbled circuits permit parties to jointly calculate a function without showing their inputs to the other, which is important for privacy-preserving protocols.

Oded Goldreich's book [18] is a foundation in cryptography. It provides careful definitions and proofs, establishing the theoretical groundwork of modern cryptographic techniques, including zero-knowledge proofs.

In this paper [10], the concept of Private Information Retrieval (PIR) is introduced, which allows a user to recover an item from a database without disclosing which item is being recovered. The authors present several PIR protocols with different efficiency and privacy characteristics.

Here, [1] on the Conference on Computer and Communications Security, 2016. Here, the work combines differential privacy into deep learning models, demonstrating how to train neural networks that provide very strong privacy guarantees for the training data. It also shows the practical techniques and experimental results, showing that privacy-preserving deep learning is practicable.

In this paper [4] the authors propose a new secure model for federated learning, which makes sure the individual updates from local devices remain private while allowing accurate aggregation of these updates. The system is designed to be efficient and sensible, making it suitable for real-world federated learning applications.

5 Result and Experiment

Here in our research lab, the models were put into practice and trained on a PC with the specified setup. CPU: Intel Core 7th Generation, RAM: 64 GB, SSD: 256 GB, HDD: 2 TB, Graphics: Radeon HD 4800 16 GB. The models given were at first executed then trained by Python as the main programming language and popular machine learning Scikit-learn which is the framework, etc. Federated learning framework.

Accuracy is regarded the most commonly used metric, rather than the efficiency of any model. However, accuracy alone is not sufficient for us to completely assess a model's flexibility, and it can occasionally be misleading (Abadi et al., 2021). As a result, the writers evaluated the given model, showing accuracy in addition to parameters such as recall, precision, and F1-score. We need to take into account the variables like False Negatives (FN), False Positives (FP), and True Positives (TP), in order to calculate these metrics as well as True Negatives (TN). In the perspective of multi-class classification, a True Positive happens generally when the classifier correctly predicts a test metrics that belongs to class X. Test example of class X that is mistakenly projected to be another class Y is known as a False Negative event. Whereas, when another class is mistakenly shows to class X, this results in a False Positive event, which is shown in the previous section. In line with this, True Negative denotes an instance in which a test sample from class Y is accurately predicted to not be a member of class X. Equations (1)–(4) offer the mathematical equations needed to calculate these measures. The percentage of accurate predictions a model makes is measured by its accuracy. It's an easy method of evaluating a model's performance.

$$\text{Accuracy} = \frac{\text{True Positive} + \text{True Negative}}{\text{Total Number of Predictions}} \tag{1}$$

$$\text{Precision} = \frac{\text{True Positive}}{\text{True Positive} + \text{False Positive}} \tag{2}$$

$$\text{Recall} = \frac{\text{True Positive}}{\text{True Positive} + \text{False Negative}} \tag{3}$$

$$\text{F1 Score} = \frac{2 \times \text{Precision} \times \text{Recall}}{\text{Precision} + \text{Recall}} \tag{4}$$

A model's recall measures how well it can locate and accurately forecast each positive event in a dataset while reducing false negatives. The F1-score strikes a compromise between accurate and thorough classification (Table 3).

A model's recall can be measured on how well it can locate and accurately think of each positive event in a dataset while making the reduction of the false negatives.

The F1-score strikes a compromise between accurate and thorough classification. Our experiment focused on using the Fed AVG algorithm across multiple clients over 20 epochs. Among every other client, the best performer was seen on client 3, particularly at epoch 3 with a 256×256 image size, achieving a global test loss of 0.4255 and a global test accuracy of 0.8016. The metrics improved gradually, reaching the best global test accuracy of 0.8649 at epoch 17, indicating the efficacy of the federated learning approach and the CNN model configuration. Client-specific metrics provided further insights. For example, Client 0's training data showed an accuracy which is of 88.26, a precision of 81.11, also recall of 84.53, and an F1-score of 2726. Its test data revealed an accuracy of 78.90, a precision of 73.91, a recall of 76.32, and an F1-score of 663. Besides, Client 1 demonstrated higher performance, with training data showing an accuracy of 91.45, precision of 94.93, recall of 94.93, and an F1-score of 5803. The test data for Client 1 reflected an accuracy of 88.55, precision of 91.08, recall of 89.80, and an F1-score of 1469.

Table 3 Evaluating metrics using federated models over test and training data set

Fed algorithm	Model	Train data				Test data			
		Accuracy	Precision	Recall	F1-score	Accuracy	Precision	Recall	F1-score
FedAVG	FED-CNN	90.51%	90.43%	90.51%	90.40%	85.74%	85.55%	85.74%	85.61%

After all, through our experiment it got highlighted in the significant improvement of Client 3 at epoch 3, making it the best of all the model configurations with a 256×256 image size. The metrics consistently improved up to epoch 17, showing us the benefits of the federated learning approach. For future experiments, it is recommended to use Client 3's model configuration, evaluate up to 3 epochs with a 256×256 image size, and continuously adjust hyperparameters to enhance model performance.

6 Hyperparameter Optimization

Hyperparameters play a significantly important role in the case of the training of ML and also in the guiding of the learning process. These parameters include various factors as for example the design of the algorithm type, model specifications, rate of learning, number of epochs, batch size, and optimization. Not similar to the parameters learned from the given data, hyperparameters are already set before the training process. Learning rate means the rate at which the machine learns. A commonly used starting point which is used in here is 0.001, which often works well through various tasks among others. While a higher learning rate means learning at a vast rate, it risks overshooting the optimal value. Just like that, the learning rate which is lower means a more static but slower curve and IOT takes more time. Epochs help to show the number of complete passes through the given training dataset. For the given experiment, 20 epochs were found to be optimal values, providing sufficient training without leading to overfitting. Batch size, another important hyperparameter, refers to the number of data samples that are processed through one iteration, which we do in 1 running. A batch size of 64 was shown to be effective, which was balancing the memory smoothness and training speed. The dimension of the image is generally used in pixels, which is the standard across all the sections. The best image size was found to be 256×256 pixels.

In this experiment, we tested various iterations of epochs, batch sizes, and image dimensions. The final and the best configuration found so far was identified as 20 epochs, a batch that has a size of 64, and image dimensions consisting of 256×256 pixels. The performance metrics showed an improvement over time, with loss reducing from 0.4394 to 0.3483 and accuracy increased from the point of 0.8011–0.8574 over the 20 epochs.

7 Complexity Analysis

The magnitude of the provided input data determines an algorithm's complexity, which begins with the amount of time and space requirements necessary to solve a fixed problem. Such complexity in the framework of federated learning mainly depends on the sharing and combining of model updates among many devices. For

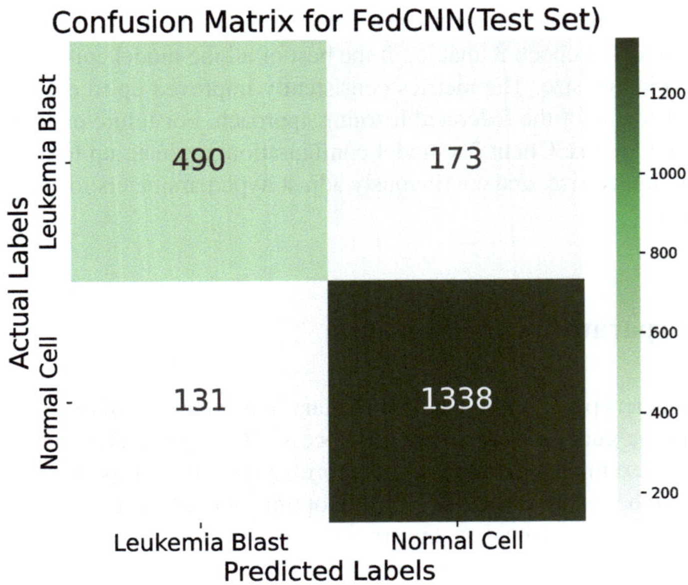

Fig. 2 Confusion matrices for the test set

example, the computing burden shows an increase of linearly with the number of devices in the Fed Avg method. The extra expenses differ based on the aggregation and optimization strategies used, as well as the proximal terms' level of complexity (Fig. 2).

Regarding specific performance metrics, the algorithm exhibited varying results: CPU times totaled 4 h, 12 min, and 18 s, while wall time amounted to 1 h, 7 min, and 7 s. In recent comparisons among different models, CNNs have demonstrated notable efficiency by completing tasks in less time compared to other models. Conversely, MobileNetV2, due to its constraints in handling image sizes, recorded the longest duration for processing. An important advantage of this system is its capability to perform detection directly on the user's device, eliminating the need to transmit data to remote servers. This not only ensures accurate results but also maintains smooth operation without affecting the performance of the user's device.

The confusion matrices reveal that the Fed CNN model has learned to classify the training data effectively, with higher correct classification rates and lower misclassification rates in the training set compared to the test set. Here the model demonstrates that on the test set, the model correctly classified 62.76% of Normal Cells and 22.98% of Leukemia Blasts, with misclassification rates of 8.11% for Leukemia Blasts as Normal Cells and 6.14% for Normal Cells as Leukemia Blasts. On the train set, the model showed higher accuracy when it correctly classified 64.59% of Normal Cells and 25.92% of Leukemia Blasts, with lower misclassification rates of 6.04% and 3.45% each and only each. These results show us that while the model performs well on the training data, it does not do much on the test data. That is, there is a slight drop in performance on the test data, which shows us that the model might be overfitting to the training set and may not generalize as effectively to new, unseen data. However, the slight drop in performance on the test set suggests that the model may be overfitting to the training data, as it does not generalize as effectively to unseen data. This discrepancy underscores the importance of further refining the model or employing regularization techniques to enhance its generalization capability.

Additionally, we can also see the performance metrics for CPU and wall time were recorded for the ongoing process, demonstrating the system's efficiency in a different context. The CPU times needed for the entire process take a total of 87.3 ms, while the wall time is 288 ms. These metrics show us the efficiency of the computational process, giving us the evaluation of the performance of different algorithms and systems. This discrepancy between training and test set performance underscores the importance of further refining the model or employing regularization techniques to enhance its generalization capability.

8 Conclusion and Future Work

Here, we proposed a framework leveraging federated learning techniques to detect leukemia from medical images while ensuring data privacy and confidentiality. The approach addresses the inherent challenges in distributed machine learning in the healthcare domain, particularly the need to preserve the privacy of medical data while enabling collaborative model training.

The implementation of the Fed AVG algorithm shows the feasibility and effectiveness of FL. Through multiple communication rounds and the use of Fed Avg for local model updates, the global model improves by iterative aggregating knowledge from diverse local datasets without sharing raw data. The experimental results show that federated learning achieves high accuracy, precision, recall, and F1-score on both training and test datasets, showing the major potential for real-world applications in medical image analysis. Our comparison of privacy-preserving techniques underscores the advantage of FL, particularly in maintaining high privacy preservation with relatively low communication complexity. The results show that federated learning can be a trustable option for sensitive data scenarios, outperforming other techniques keeping in mind information loss, accuracy, utility, and complexity. Our model, using a CNN model and hyperparameter optimization, further shows the effectiveness of the federated approach. We identify the optimal configurations, such as 20 epochs, batch size of 64, and image dimensions of 256×256 pixels. We can see that our model has improved a lot and is giving the most efficient outcome on this. Future research can build on this work by working on several key factors for further enhancement of the framework and its applications. To minimize the overfitting and improve the generalization for unseen data, future work should advance regularization techniques such as dropout, L1/L2 regularization, and data augmentation. Day by day, federated learning is becoming very much important. In future work, the researchers should focus on making the best communication protocols and aggregation algorithms to handle larger networks of clients. While FL provides privacy, incorporating additional security measures as for example the privacy and secure multi-party calculation can further enhance the privacy guarantees of the framework. In the medical domain, model interpretability is critical for gaining trust and adoption by healthcare professionals. Future research should explore methods to improve the interpretability and explainability of federated learning models.

References

1. M. Abadi, A. Chu, I. Goodfellow, H.B. McMahan, I. Mironov, K. Talwar, L. Zhang, Deep learning with differential privacy, in *Proceedings of the 2016 ACM SIGSAC Conference on Computer and Communications Security* (2016), pp. 308–318
2. F. Akram, I.U. Haq, S. Sahreen, N. Nasir, W. Naseem, M. Imitaz, A. Aqeel, Retracted: Crispr/cas9: a revolutionary genome editing tool for human cancers treatment. Technol. Cancer Res. & Treatment **21**, 15330338221132078 (2022)
3. M. Bellare, V.T. Hoang, P. Rogaway, Foundations of garbled circuits, in *Proceedings of the 2012 ACM Conference on Computer and Communications Security* (2012), pp. 784–796
4. K. Bonawitz, V. Ivanov, B. Kreuter, A. Marcedone, H.B. McMahan, S. Patel, D. Ramage, A. Segal, K. Seth, Practical secure aggregation for privacy-preserving machine learning, in *proceedings of the 2017 ACM SIGSAC Conference on Computer and Communications Security* (2017), pp. 1175–1191
5. M. Boyiadzis, P. Desai, N. Daskalakis, W. Donnellan, L. Ferrante, J.D. Goldberg, M.R. Grunwald, C. Guttke, X. Li, J.A. Perez-Simon et al., First-in-human study of jnj-63709178, a cd123/cd3 targeting antibody, in relapsed/refractory acute myeloid leukemia. Clin. Transl. Sci. **16**, 429–435 (2023)
6. H. Butt, A. Khalid, M. Yaqoob, Polymorphismm of bcl2 gene in acute lymphoblastic leukemia pastients in pakistan and screening of phytochemicals to overcome its expression. Bio Med. (Aligarh) **10** (2017)
7. B.Z. Carter, W. Tao, P.Y. Mak, L.B. Ostermann, D. Mak, G. McGeehan, P. Ordentlich, M. Andreeff, *Menin inhibition decreases Bcl-2 and synergizes with venetoclax in NPM1/FLT3-mutated AML*. Blood. The J. Amer. Soc. Hematol. **138**, 1637–1641 (2021). https://doi.org/10.1182/blood.2021011917
8. S. Chantziara, J. Musoro, A.C. Rowsell, C. Sleurs, C. Coens, M. Pe, S. Suciu, M. Kicinski, P. Missotten, E. Vandecruys et al., Quality of life of long-term childhood acute lymphoblastic leukemia survivors: comparison with healthy controls. Psychooncology **31**, 2159–2168 (2022)
9. W.A. Chen, T.G. Williams, L. So, N. Drew, J. Fang, P. Ochoa, N. Nguyen, Y. Jawhar, J. Otiji, P.J. Duerksen-Hughes et al., Duocarmycin sa reduces proliferation and increases apoptosis in acute myeloid leukemia cells in vitro. Int. J. Mol. Sci. **25**, 4342 (2024)
10. B. Chor, E. Kushilevitz, O. Goldreich, M. Sudan, Private information retrieval. J. ACM (JACM) **45**, 965–981 (1998)
11. T. Cluzeau, M. Sebert, R. Rahmé, S. Cuzzubbo, J. Lehmann-Che, I. Madelaine, P. Peterlin, B. Bève, H. Attalah, F. Chermat et al., Eprenetapopt plus azacitidine in tp53-mutated myelodysplastic syndromes and acute myeloid leukemia: a phase ii study by the groupe francophone des myélodysplasies (gfm). J. Clin. Oncol. **39**, 1575–1583 (2021)
12. C.D. DiNardo, B. A. Jonas, V. Pullarkat et al. *Azacitidine and venetoclax in previously untreated acute myeloid leukemia*. The New England Journal of Medicine, **383**(7), 617–629 (2020). https://doi.org/10.1056/NEJMoa2012971
13. N.G. Daver, P. Montesinos, A.M. Aribi, G. Martinelli, E.S. Wang, J.K. Altman, G.J. Roboz, P.W. Burke, R.B. Walter, K. Begna, et al., A phase 1b/2 study of pivekimab sunirine (pvek,

imgn632) in combination with venetoclax/azacitidine or magrolimab for patients with cd123-positive acute myeloid leukemia (aml) (2023)
14. T. Dharani, S. Hariprasath, Diagnosis of leukemia and its types using digital image processing techniques, in *2018 3rd International Conference on Communication and Electronics Systems (ICCES)* (IEEE, 2018), pp. 275–279
15. B.S. Diba, J.D. Plabon, M.M. Rahman, D. Mistry, A.K. Saha, M. Mridha, Explainable federated learning for privacy-preserving bangla sign language detection. Eng. Appl. Artif. Intell. **134**, 108657 (2024)
16. C. Dwork, Differential privacy, in *International Colloquium on Automata, Languages, and Programming* (Springer, 2006), pp. 1–12
17. C. Gentry, *A Fully Homomorphic Encryption Scheme*. (Stanford University, 2009)
18. (weizmann Institute Of Science Goldreich (Israel)), O., *Foundations of Cryptography:*, vol. 1 (Cambridge University Press, Basic Tools, 2006)
19. P. Kairouz, H.B. McMahan, B. Avent, A. Bellet, M. Bennis, A.N. Bhagoji, K. Bonawitz, Z. Charles, G. Cormode, R. Cummings, et al., Advances and open problems in federated learning. Foundations and Trends® Mach. Learn. **14**, 1–210 (2021)
20. M.G. Kibriya, D. Mistry, D. Mistry, M.T. Rahman, S.H. Ovi, A. Tabassum, S. Hossain, Privacy-preserving vision-based detection of pox diseases using federated learning, in *Data-Driven Clinical Decision-Making Using Deep Learning in Imaging*. (Springer Nature Singapore, Singapore, 2024), pp.105–122
21. M. Lou, H. Ying, X. Liu, H.Y. Zhou, Y. Zhang, Y. Yu, Sdr-former: A siamese dual-resolution transformer for liver lesion classification using 3d multi-phase imaging (2024). arXiv:2402.17246
22. D. Mistry, J.D. Plabon, B.S. Diba, S. Mukta, M.F. Mridha, Federated learning-based architecture for personalized next emoji prediction for social media comments. IEEE Access **1–1** (2024). https://doi.org/10.1109/ACCESS.2024.3448470
23. D. Mistry, M.R. Tonmoy, M.S. Anower, A.T. Hasan, Federated transfer learning for vision-based fall detection, in *International Conference on Big Data, IoT and Machine Learning*. (Springer, 2023), pp. 961–975
24. C. Nollmann, W. Moskorz, C. Wimmenauer, P.S. Jäger, R.P. Cadeddu, J. Timm, T. Heinzel, R. Haas, Characterization of cd34+ cells from patients with acute myeloid leukemia (aml) and myelodysplastic syndromes (mds) using a t-distributed stochastic neighbor embedding (t-sne) protocol. Cancers **16**, 1320 (2024)
25. J.M. Nørgaard, L.S Friis, J.S. Kristensen, M.T. Severinsen, I. Mølle, C.W. Marcher, P. Møller, C. Schoellkopf, O.J. Nielsen, B.S. Preiss, M.K. Andersen, E. Kjeldsen, B.C. Medeiros, L.S.G. Østgård, *Addressing the room for improvement in management of acute promyelocytic leukemia*. European Journal of Haematology, **102**(6), 479–485 (2019). https://doi.org/10.1111/ejh.13229
26. E. Orgel, T.B. Alexander, B.L. Wood, S.B. Kahwash, M. Devidas, Y. Dai, T.A. Alonzo, C.G. Mullighan, H. Inaba, S.P. Hunger et al., Mixed-phenotype acute leukemia: a cohort and consensus research strategy from the children's oncology group acute leukemia of ambiguous lineage task force. Cancer **126**, 593–601 (2020)
27. M.M. Patnaik, T.I. Mughal, C. Brooks, R. Lindsay, N. Pemmaraju, Targeting cd123 in hematologic malignancies: identifying suitable patients for targeted therapy. Leukemia & Lymphoma **62**, 2568–2586 (2021)
28. N. Santoro, P. Salutari, M. Di Ianni, A. Marra, Precision medicine approaches in acute myeloid leukemia with adverse genetics. Int. J. Mol. Sci. **25**, 4259 (2024)
29. K. Seipel, H. Mandhair, U. Bacher, T. Pabst, Flt3 and irak4 inhibitor emavusertib in combination with bh3-mimetics in the treatment of acute myeloid leukemia. Curr. Issues Mol. Biol. **46**, 2946–2960 (2024)
30. S.M. Seyedzade, S. Mirzakuchaki, R.E. Atani, A novel image encryption algorithm based on hash function, in *2010 6th Iranian Conference on Machine Vision and Image Processing* (2010), pp. 1–6
31. M.S.H. Shovon, Z. Sultana, J. Shin, M.A. Hamid, D. Mistry, Incorporating residual connections into a multi-channel cnn for lung cancer detection in digital pathology, in *Data-Driven Clinical*

Decision-Making Using Deep Learning in Imaging (Springer Nature Singapore, Singapore, 2024), p. 189
32. A. Tabassum, S.H. Ovi, S. Hossain, M.R. Tonmoy, M.S.H. Shovon, M.R. Hussein, D. Mistry, Privacy preserving breast cancer prediction with mammography images using federated learning, in *Data-Driven Clinical Decision-Making Using Deep Learning in Imaging*. (Springer Nature Singapore, Singapore, 2024), pp.227–245
33. I. Vincent, K.R. Kwon, S.H. Lee, K.S. Moon, Acute lymphoid leukemia classification using two-step neural network classifier, in *2015 21st Korea-Japan Joint Workshop on Frontiers of Computer Vision (FCV)* (IEEE, 2015), pp. 1–4
34. R. Walewska, N. Parry-Jones, T.A. Eyre, G. Follows, N. Martinez-Calle, H. McCarthy, H. Parry, P.E. Patten, J.C. Riches, P. Hillmen et al., Guideline for the treatment of chronic lymphocytic leukaemia. Br. J. Haematol. **197**, 544–57 (2022)
35. S.J. Wu, S.Y. Huang, C.T. Lin, Y.J. Lin, C.J. Chang, H.F. Tien, The incidence of chronic lymphocytic leukemia in taiwan, 1986–2005: a distinct increasing trend with birth-cohort effect. Blood. J. Amer. Soc. Hematol. **116**, 4430–4435 (2010)

Secure Collaborative Learning for CSV-Based Ovarian Cancer Diagnosis: A Federated Approach

Bidita Sarkar Diba, Md. Arafat Kabir, Tasnim Jahin Mowla, Hanif Bhuiyan, and Durjoy Mistry

Abstract Ovarian cancer, one of the most dangerous malignancies of the female reproductive system, demands accurate and early diagnosis. This study proposes a federated learning system for collaborative image-based ovarian cancer diagnosis, enabling multiple healthcare institutions to train a robust diagnostic model without compromising patient data privacy. By utilizing advanced techniques like differential privacy and secure aggregation, the system addresses challenges such as data heterogeneity and communication efficiency. Experimental results demonstrate that the proposed model achieves a test accuracy of 95.71% on a dataset of over 349 raw data points, performing comparably to centralized models while maintaining stringent privacy standards. This work highlights the potential of federated learning to revolutionize medical research and improve diagnostic outcomes for ovarian cancer.

Keywords Ovarian cancer · Machine learning · Federated learning · Privacy preservation · Logistic regression

B. S. Diba · Md. A. Kabir · T. J. Mowla · D. Mistry (✉)
Department of CSE, University of Asia Pacific, Dhaka, Bangladesh
e-mail: durjoy@uap-bd.edu

B. S. Diba
e-mail: bidita@uap-bd.edu

Md. A. Kabir
e-mail: 21201146@uap-bd.edu

T. J. Mowla
e-mail: 21101069@uap-bd.ed

H. Bhuiyan
Monash Data Futures Institute Monash University, Melbourne, VIC, Australia
e-mail: hanif.bhuiyan@monash.edu

M. F. Mridha and N. Dey (eds.), *Federated Learning in Health Care Technology*, Studies in Computational Intelligence 1216, https://doi.org/10.1007/978-981-96-8353-6_7

1 Introduction

The introduction of sophisticated medical imaging technology has greatly improved the identification and management of ovarian cancer, which is a major contributor to female cancer mortality globally. In computer vision, cancer detection, particularly ovarian cancer, is a challenging problem. Even though ovarian cancer detection is the subject of several research [1–4]. In developed nations, ovarian cancer ranks fifth in cancer-related deaths among women and sixth in overall cancer incidence [5]. Globocan's 2022 forecasts indicate that by 2050, 503,448 women worldwide will get an ovarian cancer diagnosis, an increase of approximately 55%. It is anticipated that 350,956 women will lose their lives to ovarian cancer this year, an increase of over 70% from 2022. Both nonepithelial and epithelial ovarian cancers are classified as ovarian cancers. More than 95% of ovarian cancer cases are of the epithelial type, with the remaining 5% being nonepithelial malignancies (such as sex-cord stromal, small cell, and germ cell ovarian cancers). Histologic categorization is used to categorize epithelial ovarian cancers into subtypes. These subtypes include endometrioid, high-grade serous, low-grade serous, clear cell, and mucinous ovarian cancer, and their differences may affect patient outcomes, diagnostic evaluation, and treatment [6]. The ovary's epithelial cells are the source of most ovarian cancers [5]. Types of ovarian cancer:

- Epithelial: Epithelial ovarian cancer is the most common kind. Primary peritoneal and fallopian tube cancers are treated in a manner similar to epithelial ovarian cancers.
- Germ cell: Ovarian germ cell tumors are uncommon and often afflict girls and young women in their early thirties. Ovarian teratomas are the most prevalent kind.
- Sex cord-stromal: Ovarian stromal tumors of the sex cord may be benign (non-cancerous) or malignant (cancerous). Tumors with granulosa cells are the most prevalent kind.
- Borderline: Aberrant cells that develop in the tissue surrounding the ovary are known as borderline ovarian tumors. Surgery is often used to cure them since they are not cancer.

Numerous algorithms related to machine learning (ML) and deep learning (DL), such as DT, ANN, SVM, CNN, RF, KNN, GBM, RF, ResNet-50, Dense Net, DCNN, GARSINFIS, VGNet, AlexNet, and FCNN, among others, have been used in a range of cancer detection research projects. Remember that traditional ML approaches often need to collect data on a solitary server or desktop. The data must first be gathered and processed on a central computer before it can be given back to the devices, which is one drawback of this system. Addressing privacy concerns is essential since people's medical records include sensitive information [7]. The fact that the raw data is sent to a central computer raises legitimate privacy concerns. Machine learning and deep learning models, especially those trained on sensitive datasets, are susceptible to model inversion assaults. These attacks exploit the model's outputs to infer sensitive information from the training set by utilizing the model's predictions to recreate words or images. Model inversion attacks provide a substantial threat to individuals' right to privacy, in particular for computing systems that are dependent on surveillance, medical diagnosis, or biometric identification mechanisms.

In this research, an innovative method for the identification of ovarian cancer based on a CSV file is referred to as Secure Collaborative Learning (SCL). The goal of this approach is to overcome the challenges that have been presented. We use the federated learning paradigm, which is a decentralized method to machine learning that safeguards the confidentiality of data and makes it possible for several institutions to work together to train models by collaborating. Through the distribution of model training to local datasets without the exchange of raw data, federated learning contributes to the maintenance of compliance with legal frameworks like as the General Data Protection Regulation (GDPR)[7] and the Health Insurance Portability and Accountability Act (HIPAA). Within this paper, we provide a comprehensive explanation of our federated approach to the diagnosis of ovarian cancer. This description includes details on the method's architecture, workflow, and the fundamental concepts that constitute it. We demonstrate the feasibility and efficiency of our technique by comparing its outcomes to those of centralized learning systems. This demonstration is accomplished via exhaustive testing. In addition to this, we evaluate the privacy and security safeguards that our federated system provides, highlighting the fact that it has the potential to facilitate collaborative research and innovation in the field of medical imaging without compromising the confidentiality of the data.

Federated learning improves the robustness and generalizability of diagnostic models during the diagnosis of ovarian cancer by facilitating the integration of heterogeneous imaging datasets from several institutions. The overall result of this cooperative endeavor is better patient care and diagnostic accuracy by using the data strengths of each participating institution while maintaining anonymity.

The research endeavor has yielded the following significant contributions:

- It is proposed to develop an architecture that prioritizes safeguarding user privacy. The stated purpose of this technology is to recognize characters from medical imaging datasets.
- accuracy 95.71%
- Using the most extensive medical imaging dataset currently accessible, the authors trained and evaluated their model.
- Using SHAP (Shapley Additive explanations), the writers explore and classify the predictions made by these models.

This work seeks to enhance the existing knowledge of federated learning in medical imaging and showcase its capacity to revolutionize cancer detection. Our objective is to promote safe and efficient cooperation across healthcare facilities, with the goal of fully using innovative technology to improve patient outcomes and progress in the area of cancer. Overall, our work contributes to the advancement of secure and privacy-preserving machine learning techniques in healthcare, with the ultimate goal of improving early detection and treatment outcomes for ovarian cancer patients.

The rest of the paper is structured as follows: Related Work explores methods and innovations for diagnosing and managing ovarian cancer using deep learning and machine learning. Conventional Privacy-Preserving Methods address approaches to secure sensitive dataset information related to ovarian cancer patients. Methodology provides a dataset description, preprocessing steps, the proposed framework, and the application of the Fed algorithm. Result & Analysis presents experimental results, evaluation metrics, and comparisons with prior research. Conclusion and Future Work highlights the study's findings, contributions, and potential directions for further exploration.

2 Related Work

Ovarian cancer is a malignancy originating within the ovaries, integral components of the female reproductive system. Sufficient research has been conducted to identify ovarian cancer. Researchers have used various types of machine learning, deep learning algorithms, and hybrid models to identify ovarian cancer.

2.1 Literature Review

Zhuang et al. [1], demonstrated a classification model using SE–SPP–DenseNet architecture to predict platinum resistance in patients based on imaging data [1]. To address the substantial disparity in class distribution between patients with platinum resistance (97) and platinum sensitivity (192), data augmentation was used on the platinum resistance data. This included rotating the pictures by 90° to expand the dataset size. In all, the dataset was partitioned into validation, test, and training sets in a ratio of 0.1:0.1:0.8, resulting in 3400, 430, and 430 samples, respectively. DenseNet121 served as the backbone of the convolutional neural network (CNN) architecture, and an SE Block and SPP Layer were incorporated for prediction. The SE Block, positioned between convolutional layers, facilitated feature recalibration, while the SPP Layer, replacing the pooling layer in DenseNet, enabled multiscale feature extraction. Experimental images were resized to 224 × 224 pixels. Additionally, ResNet18 with SE Block and SPP Layer was employed for medical-image classification, enhancing feature extraction while preventing gradient dispersion. The Swin Transformer model was also utilized, employing a hierarchical construction approach and a sliding window mechanism for feature extraction from multiple windows. Enhancements included a Patch Partition Block for downsampling images and a Patch Merging Layer for feature map adjustment. The Swin Transformer block utilized windowed multihead self-attention to classify images effectively. The accuracy of the model was 92.6%.

In their 2023 study, Heartlin Maria et al. implemented a novel hybrid model consisting of two deep learning networks; the YOLO v5 and U-Net models are used for the identification and separation of ovarian tumors. The YOLO v5 model was used for the purpose of detecting Regions of Interest (ROI), while the U-Net model was used to conduct semantic segmentation on the identified ROIs [8]. The datasets utilized in this study were obtained from the RSNA (Radiological Society of North America) and TCIA (The Cancer Imaging Archive). Preprocessing of the acquired datasets involved image resizing, denoising, and augmentation techniques. The accuracy of the model achieved 98%. Although the proposed hybrid deep learning approach effectively detects and segments ovarian tumors, it does not address the localization of the tumors or any specific subtypes.

A genetic algorithm-based rough set clustering method called Genetic Algorithm and Rough Set Incorporated Neural Fuzzy Inference System (GARSINFIS), a six-layered, feed-forward, partially connected architecture, was proposed by Di Yang et al. [9]. The method combines genetic algorithms with rough set theory. Fashions can identify the early stages of cancer correctly, but they are unable to detect the disease after it has progressed to a malignant stage.

Ahamad et al. [10] analyzed the accuracy of various machine learning algorithms together with Logistic Regression (LR), Decision Tree (DT), Random Forest (RF), Light Gradient Boosting Machine (LGBM), Support Vector Machine (SVM), Extreme Gradient Boosting Machine (XGB), and Gradient Boosting Machine (GBM) individually for the prediction of ovarian cancer [10]. The dataset, comprising information from 349 patients, was obtained from the Third Affiliated Hospital of Soochow University. Among these patients, there were 171 diagnosed with ovarian cancer, and 178 with benign ovarian tumors, and the dataset included 49 features. The dataset is divided into three subgroups: Blood Routine Test, General Chemistry, and Tumor Marker. Different classifiers were employed to assess the accuracy of the dataset, including Decision Trees (DT) with an accuracy of 87%, XGBoost with 80% accuracy, Support Vector Machine (SVM) and Extreme Learning Machine (ELM) both with an accuracy of 87%, Random Forest (RF), Gradient Boosting Machine (GBM), and LightGBM (LGBM) all with an accuracy of 88%, and RF with 91% accuracy. Only 106 patients with OC marker features were utilized to achieve the highest accuracy.

Wen et al. [11] utilized a texture exploration approach to appraise the structural variations in the extracellular matrix (ECM) seen in second-harmonic generation (SHG) pictures of high-grade ovarian serous carcinoma and normal ovarian stroma [11]. ROC curve study showed that they were able to attain a noteworthy accuracy level of 97% in differentiating between high-grade malignant tissue and normal ovarian tissue by optimizing the texton count and closest neighbors (NNs). Their technique has the potential to be extended to other situations, such as low-grade and borderline tumors. Their application concentrated on the discrete issue of differentiating between normal and high-grade serous cancers.

Hybrid Particle Genetic Swarm Optimization (PGSO) to optimize the characteristic selection and Multi-Class Support Vector Machine (SVM) is demonstrated by Yasodha and Ananthanarayanan [12] for classifying different stages of ovarian cancer [12]. This research used a multiclass Support Vector Machine (SVM), Artificial Neural Network (ANN), and Naive Bayes classifier to do classification. Two datasets were used, one acquired via the TCGA site and the other from the Singapore National University Hospital (NUH). Dataset 2 obtained the maximum accuracy of 96%. The authors have offered a little discussion about the applicability of the suggested knowledge base system.

In 2015, The authors(A Enshaei, C N Robson, R J Edmondson)developed a model using an artificial neural network (ANN) algorithm that can predict epithelial ovarian cancer(93%) and area 0.74 under the curve(AUC) which exhibits superior performance relative to logistic regression [13]. The model was trained by 668 clinical data, but the features of the data are hidden.

BenTaieb et al. [14] presented two model classifiers using K-means CSP features in conjunction with a latent structured SVM model, as well as a classifier employing DN features with Fisher encoding and linear SVM [14]. The suggested approaches depend on poorly supervised learning and only use whole slide image (WSI) labels, which signal the existence but not the precise location of a specific ovarian cancer inside the imaged tissue.

An accuracy rate of 78.2% was attained using a Deep Convolutional Neural Network (DCNN) that was created from the AlexNet model. This accuracy was determined by analyzing a dataset of 85 images exclusively. The model was developed by Miao Wu, Chuanbo Yan, Huiqiang Liu, and Qian Liu in the year 2018 [15]. Certain photos of ovarian cancer are incorrectly categorised, especially those that lack distinct cell shapes. The model may not be appropriate for all categories of ovarian cancer photos, particularly those exhibiting cell overlap or inadequate clarity.

In 2019, Lei Zhang, Jian Huang, and Li Liu proposed the use of a fine-tuned GoogLeNet neural network and the low-level ULBP model for the early diagnosis of ovarian cancer using a color ultrasound detection system [16]. This study included 357 cases of malignant tumors, 71 cases of benign tumors, 277 samples of ovarian cancer, 299 samples of benign cysts, and 824 photos showing normal ovarian structure. The advanced deep learning approach proposed in this study has the potential to improve recognition accuracy. Nevertheless, if the image quality is inadequate, the technique outlined in this study may be ineffective.

Md. Ali Hossain et al. (2019) presented a model trained using the TCGA-OV dataset employing Cox proportional hazard regression analysis [17]. The study utilized gene expression data obtained from RNA sequencing (RNAseq) experiments, which are subject to inherent limitations and assumptions, including the necessity for standardizing transformations and the possibility of skewness in gene expression distributions.

Akazawa and Hashimoto [18] delineated various machine learning algorithms and their respective accuracies [18]. In this experimental investigation, the authors evaluated the performance of each algorithm using 16, 10, and 6 features. The maximum achieved accuracy was 80%(XGBoost) when utilizing 16 features.

In their 2021 study, Rania M. Ghoniem et al. introduced a hybrid evolutionary deep learning model utilizing multi-modal data for the diagnosis of ovarian cancer. The proposed model involves an improved Long Short-Term Memory (LSTM) architecture optimized using the Ant Lion Optimizer (ALO) algorithm for processing longitudinal gene data [19]. The study utilizes the TCGA-OV, BRCA, and LUSC datasets. The dataset comprises gene expression and copy number variant data for 587 patients diagnosed with ovarian cancer (OC). Additionally, each patient's records include multiple pathological images. The gene expression and copy variant data are represented in one dimension, constituting one modality, while the pathological images, which are in color, represent another modality. The accuracy of the model was 97%.

Kasture et al. [20] explained a novel deep convolutional neural network (DCNN) architecture comprising 8 Convolutional Layers, 4 Maxpooling Layers, 3 Fully Connected Layers, an input image layer, and a Fully Connected Output Layer [20]. The model incorporates eight parameters including Learning Rate, Cost function, Optimizer, Epoch Number, Batch Size, Dropout (Convolution Layer), Dropout (Dense Layer), and Activation function. The accuracy achieved by the model is 83.93%. The model was trained using a dataset divided into four classes: Clear Cell, Endometrioid, Mucinous, Non-Cancerous, and Serous.

In 2022, A machine learning model (ML) trained using the support vector machine (SVM) and k-nearest neighbors (KNN) techniques was reported by Nasser Taleb et al. [21]. Data from 463 patients, including both malignant and non-cancerous instances, were included in the dataset utilized in this investigation. It is composed of 48 independent variables and 1 dependent variable, which are separated into two classes: normal and ovarian cancer. The model achieves a 98% accuracy. Nevertheless, a thorough examination of the computer power needed to put the suggested paradigm into practice is absent from the study (Table 1).

2.2 *Privacy-Preserving Techniques*

Methods and procedures developed to safeguard the privacy and confidentiality of photographs are known as image privacy-preserving approaches. In this day and age of ubiquitous digital image sharing, when pictures might include private or sensitive information that shouldn't be seen by others, these methods have never been more crucial. Both academic research and industrial domains have shown great interest in several privacy-preserving approaches. Table 2, a comparative comparison of privacy-preserving strategies, was prepared by the authors.

3 Proposed Methodology

The writers will begin by offering a thorough analysis of the dataset in this part. They will then go on to explain how the suggested framework functions. They will then go into more detail on how federated learning protects privacy.

Table 1 Summary of related work

SL No	Year	Model	Accuracy	Limitations
Paper [9]	2014	GARSINFIS, C4.5, Naive, MLP, RBF, ANFIS, DENFIS	80%	Poor performance in malignant stages.
Paper [11]	2014	kh2NN, The MR8	97%	Low accuracy for low-grade and borderline tumors.
Paper [12]	2015	Multiclass SVM, ANN, Naive Bayes	96%	Limited discussion on the generalizability of the proposed knowledge base system.
Paper [13]	2015	ANN, DT, BN	93%(ANN)	Features are hidden.
Paper [22]	2016	DT	76.70%	Low accuracy.
Paper [14]	2017	CSC, DN, CNN, SVM, LSVM, SVM	90.00%	Indicates the presence but not the location of a particular ovarian carcinoma within the imaged tissue
Paper [15]	2018	DCNN based on AlexNet	78.20%	The current model may not be suitable for all types of ovarian cancer images, especially those with overlapping cells or poor clarity.
Paper [16]	2019	VGNet, AlexNet, FCNN, CNN, GoogLeNet	99%(Proposed)	A small number of samples were used.
Paper [17]	2019	Cox proportional hazard regression analysis	N/A	Not mention the accuracy.
Paper [23]	2020	Cox proportional hazard regression analysis, LASSO	N/A	Not mention the accuracy
Paper [18]	2020	RF, SVM, NB, LG, XGBoost	80.0%(XGBoost)	A small number of features are used in the algorithm.
Paper [19]	2021	LSTM, CNN(ResNet34, VGG16)	97%	The research paper does not address the computational complexity.
Paper [20]	2021	DCNN	84%	Use few numbers of models
Paper [21]	2022	KNN, SVM, Dense Net	98%	Missing resources
Paper [10]	2022	RF, SVM, DT, XGBoost, LR, GBM, LGBM	91%(RF)	Only applicable for clinical data.
Paper [8]	2023	Hybrid (YOLO v5 and U-Net)	98.00%	Overfitting
Paper [4]	2023	ANN, LDA AND SVM	98.00(SVM)	A small amount of data is used for testing.
Paper [3]	2023	RF, AdaBoosting, XGBoosting	91.37%	Applicable for only benign and malignant tumors.
Paper [2]	2024	ResNet-50	90.00%	The model does not perform well for borderline OC.
Paper [1]	2024	Resnet18, SE Block, SPPLayer, DenseNet, Swin Transformer	92.60%	Limited training data.

Table 2 Privacy-Preserving comparison of different technique [24, 25]

Techniques	Accuracy	Privacy	Information loss	Communication complexity	Computational complexity	Utility
Homomorphic Encryption	High	Medium	Medium	High	High	High
ConvMixer	High	Medium	Medium	Low	Low	High
Differential Privacy	Low	High	Low	Low	Low	Low
Double Random Phase Encoding (DRPE)	Low	Low	High	High	High	Medium
Secure Multi-Party Computation (MPC)	Medium	High	Low	High	High	Medium
Block-Wise Scrambled Images	High	High	Low	Low	Low	High
Cryptographic Hash Functions	Medium	Low	High	Low	Low	High
Private Set Intersection (PSI)	Medium	Medium	Medium	Low	Low	High
Federated Learning	High	High	Low	Low	Low	Medium

3.1 Dataset Description

This research utilizes a Kaggle dataset to predict ovarian cancer, comprising 349 raw data entries in CSV format. The dataset is organized into five distinct parts.

- Part 1: holds the original raw data.
- Part 2: Includes a chart of biomarkers, along with their descriptions and abbreviations, applied in the study.
- Part 3: Consists of the scattered version of the training data, excluding the biomarker CA72-4.
- Part 4: holds the raw training data.
- Part 5: holds the raw test data.

The dataset includes 47 biomarkers, with serum and blood samples used in this research. The data is divided such that 80% is used for testing and 20% is used for training.

3.2 Proposed Framework

Federated learning is a cutting-edge machine learning method that allows several decentralized devices or servers to collectively train a common model while preserving the data on the local devices. This technique mitigates privacy and data security issues by guaranteeing that raw data remains confined to local contexts and is never sent beyond. Instead, every device that is involved independently handles its own input and transmits only model changes (such as gradients) to a central server. The central server then combines these updates to enhance the overall model. Within the scope of the Federated Learning approach, data is characterized by differences in quality, variety, and quantity. These variations indicate the varied characteristics of the computing devices that are actively participating in the process. User-specific data is utilized when it comes to training objectives that are limited to the local domain.

Notable characteristics of Federated Learning

I. Data Privacy: Federated learning reduces the likelihood of data breaches and unauthorized access by storing data on local devices, which is particularly important for applications that include sensitive information.
II. Dispersed Training: Training takes place on many devices, such as smartphones, IoT devices, or edge servers, enabling the use of dispersed data without centralization.
III. Streamlined Aggregation: The central server consolidates changes to the model from individual devices, enhancing the overall model incrementally without directly accessing the data stored on each device.
IV. Bandwidth Efficiency: Minimizes the need for extensive data transfers by only exchanging model updates, which are often much less in size compared to the original data.
V. Scalability: Able to expand and adapt to a large number of devices, making it well-suited for situations with data sources spread out across a vast area.

Types of federated learning:

I. Centralized Federated Learning: Centralized federated learning is a machine learning technique in which training data is stored and processed on a single server, but numerous client devices contribute to the training process. This approach necessitates the existence of a central server to oversee the initial selection of client devices and to gather model changes throughout the training process. Communication is only established between the central server and specific edge devices. While this approach is rather simple and capable of generating precise models, the central server presents a bottleneck problem. Network failures have the potential to completely stop the whole process, emphasizing the system's reliance on the continued functioning of the central server.
II. Decentralized Federated Learning: Decentralized federated learning eliminates the need for a central server to supervise the learning process. Instead, changes to

the model are immediately transferred between edge devices that are connected to a network. The ultimate model is acquired on an edge device by consolidating the local updates from connecting edge devices. This method mitigates the danger of a single point of failure. The precision and dependability of the ultimate model are significantly influenced by the network structure and the effectiveness of communication among the edge devices.

III. Heterogeneous federated learning: Heterogeneous federated learning encompasses the use of diverse clients, such as mobile phones, desktops, or IoT (Internet of Things) devices. These devices may vary in terms of their hardware, software, computational capability, and sorts of data. HeteroFL was created as a solution to the prevalent Federated Learning approaches that make the assumption that the characteristics of the local models are similar to those of the main model. However, in reality, it occurs seldom. HeteroFL has the ability to produce a unified model for making deductions by training on numerous diverse local models.

The authors have delineated each stage of this suggested framework using a Fig. 1.

- Steps of the working model:

  I. Initialize global model: Initializing the model on the server is the first step that we take. The process is precisely the same in traditional centralized learning: we begin by initializing the model parameters, either by selecting them at random or by using a checkpoint that was previously stored.

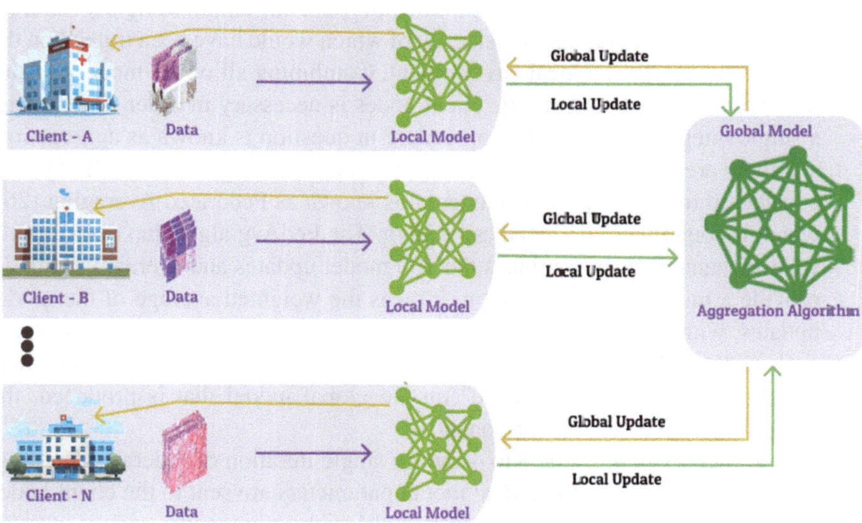

Fig. 1 Federated Learning framework with decentralized model training and central aggregation

II. Distribute the model to a number of client nodes, which are linked organizations and apparatuses: The parameters of the global model are then sent to the client nodes that are linked to the network. This is done to guarantee that all of the participating nodes begin their local training with identical model parameters. In many cases, we merely make use of a small number of the linked nodes rather than all of them. The reason for this is that the return on investment decreases as the number of client nodes that are selected increases.
III. Locally train the model using the data from each organization/device (client node): The local training process is now underway on all (chosen) client nodes as they now know the most recent version of the global model parameters. To train their own local model, they use their own local dataset. The model is merely trained for a brief period of time, not until complete convergence. This might be limited to a few steps (mini-batches) or possibly just one epoch on the local data.
IV. Send back model changes to the server. Following the completion of the local training, every client node has a slightly modified version of the model parameters that they were initially given. Due to the fact that every client node has a unique collection of samples in its local dataset, the parameters are all distinct from one another. The client nodes then communicate the model changes to the server in order to receive them. The updates to the model that they deliver may either be the whole set of model parameters or simply the gradients that were gathered during the aforementioned local training.
V. Add changes to the models together to make a new world mode: Updates to the model are sent to the server from the client nodes that have been chosen. If it chose one hundred client nodes, it would now have one hundred slightly different copies of the initial global model, each of which would have been trained on the data from one client's local environment. Combining all of the model updates that we have received from the client nodes is necessary in order for us to get a single integrated model. The procedure in question is known as aggregation, and there are a great deal of diverse approaches to taking it. The method that is considered to be the most fundamental is known as Federated Averaging [26], which is sometimes shortened as FedAvg. The FedAvg algorithm does exactly what its name suggests: it takes the 100 model updates and averages them. To provide a more exact explanation, it takes the weighted average of the model updates, with the weighting being determined by the number of instances that each client utilizes for training. For the purpose of ensuring that every data sample has the same "influence" on the global model that is produced, the weighting is an essential component.
VI. Repeat Steps 1 to 4: Steps 1 to 4 form a single iteration of federated learning. During the first phase, the global model parameters are sent to the client nodes that are participating. During the second phase, the client nodes engage in training using their own local data. During the third phase, the revised models are sent back to the server. Ultimately, during the fourth stage, the server merges the model changes to produce a fresh iteration of the global model. During a single round, each participating client node just experiences a concise training

session. Following the aggregation stage (Step 4), the model undergoes training using the data from all client nodes, although for a short duration. It is essential to repeat this training process several times to finally build a fully trained model that demonstrates great performance across all client nodes' data.

- Components of the working model:

I. Clients: These individuals are the final users who are involved in the work of prediction of ovarian cancer.
II. Global Model: The global model is crucial in orchestrating the learning process. It gathers parameters from general models provided by customers, making it easier to create a complete global model.
III. Local Model: It is the client's responsibility to train local models using their own personal datasets. The contributions that are made by a variety of local models are of great assistance in enhancing the centralized model, which is situated at the main station on a worldwide level. The global model that has been merged is then returned to the different local models so that they may continue to utilize it and continue to evolve it via iterations.
IV. Input: The structure accepts clinical CSV data as its input.
V. Output: The result of the framework is the projected occurrence of Ovarian cancer.

Participants would submit their modified model back to the centralized server and maintain their dataset in their own safe infrastructure after two rounds of local training. The central server would then combine all member inputs. Participating institutions would get the revised criteria to continue local training [27–30].

3.3 Proposed Algorithm

A gradient-based Federated Optimization (FedOpt) algorithm is a crucial technique employed in federated learning systems, where a variety of clients, such as mobile devices and edge devices, collaborate to collectively train machine learning models. In this approach, each client autonomously computes gradients of the loss function using its local data and forwards these gradients to a central server. The server then aggregates these gradients, adjusts the global model parameters accordingly, and redistributes the updated model back to the clients. This iterative process continues for a specified number of rounds, facilitating the gradual improvement of the global model while ensuring the confidentiality of each client's data.

Algorithm 1 Gradient-Based Federated Optimization (FedOpt)

Input: Learning rate η, number of clients K, number of rounds T
Initialize global model parameters $\mathbf{w}_0$
for each round $t = 0, 1, \ldots, T-1$ **do**
  Server broadcasts $\mathbf{w}_t$ to all clients
  for each client $k = 1, \ldots, K$ **in parallel do**
    Client k receives $\mathbf{w}_t$ and computes gradient $\nabla f_k(\mathbf{w}_t)$ using local data
    Client k sends gradient $\nabla f_k(\mathbf{w}_t)$ to the server
  end for
  Server aggregates gradients: $\nabla f(\mathbf{w}_t) = \frac{1}{K}\sum_{k=1}^{K} \nabla f_k(\mathbf{w}_t)$
  Server updates global model: $\mathbf{w}_{t+1} = \mathbf{w}_t - \eta \nabla f(\mathbf{w}_t)$
end for
Output: Final model parameters $\mathbf{w}_T$

Notation:

- $\mathbf{w}_t$: Global model parameters at round t
- η: Rate of the Learning
- K: Quantity of the clients
- T: Quantity of the rounds
- $\nabla f_k(\mathbf{w}_t)$: The gradient of the loss function is computed by the client k at round t.
- $\nabla f(\mathbf{w}_t)$: Aggregated gradient at round t.

3.4 *Explanation of the Algorithm 1*

The gradient-based FedOpt algorithm ensures that the model parameters are updated in a distributed manner while maintaining data privacy. Each client computes gradients using its local data and sends these gradients to the server, which aggregates them and updates the global model.

1. **Initialization:** A server initializes the global structure parameters denoted as w_0. The learning rate, represented by η, along with the specified quantity of clients denoted as K, and the total quantity of rounds, indicated by T, are defined beforehand.
2. **Broadcasting:** At the start of every round t, the server transmits the present global structure parameters w_t all customers.
3. **Local Computation:** Every client k accepts the global model parameters w_t from the server. Using its local dataset, the client computes the gradient of the loss function, denoted as $\nabla f_k(w_t)$. This process includes performing both forward and backward passes through the local data to evaluate the gradient accurately. The forward pass involves passing the data through the model to compute predictions, as the inverse pass determines the loss function's gradient in relation to the model parameters.

4. **Gradient Transmission:** After computing the local gradient, each client sends its computed gradient $\nabla f_k(\mathbf{w}_t)$ in reverse to the server.
5. **Gradient Aggregation:** The server receives the gradients from all clients and aggregates them. The aggregated gradient $\nabla f(\mathbf{w}_t)$ is computed as the average of the gradients from all clients:

$$\nabla f(\mathbf{w}_t) = \frac{1}{K} \sum_{k=1}^{K} \nabla f_k(\mathbf{w}_t)$$

6. **Model Update:** The server revises the global model parameters by utilizing the aggregated gradient from all clients along with the learning rate η. This update step is crucial as it integrates the information from every client to refine the parameters of the global model, ensuring that the model learns from the distributed data effectively.

$$\mathbf{w}_{t+1} = \mathbf{w}_t - \eta \nabla f(\mathbf{w}_t)$$

7. **Iteration:** Steps 2 to 6 are repeated for a specified count of rounds T. Each round improves the global model founded on the gradients calculated by the clients.
8. **Output:** After T rounds, the final model parameters $\mathbf{w}_T$ are output by the server.

4 Result and Experiment

In order to get a more favorable result, we carried out a number of tests using a variety of Deep Learning models. The classification of ovarian cancer was accomplished by the use of five different pre-trained models in combination with logistic regression.

4.1 Experiment Setup

At the research facility, we used a computer to train and apply the models, and we used the parameters that were given. The following is a list of the criteria that the system must meet: Graphics: NVIDIA GeForce RTX 4080 with 16GB of dedicated memory, CPU: Intel Core i7 13th Generation, RAM: 64GB, solid-state drive (SSD): 256GB, hard disc drive (HDD): 2TB, and graphics card: 256GB. A number of well-known frameworks of machine learning, including as Scikit-learn, TensorFlow, computer vision, neural networks, and Keras, have been used by the authors. Additionally, they have used the federated learning framework Flower, as well as constructed and trained the models using the programming language which is Python.

4.2 Evaluation Metrics

To measure how well our method works, we employ a variety of measures. We used a plethora of pre-trained deep learning models to examine the dataset inside a federated setting. The implementation of a random selection approach allowed for the assessment of pre-trained models. A total of 80% of the samples were used for training, with the remaining 20% going toward testing. The proposed model is assessed using a battery of metrics, including precision (P), recall (R), accuracy (A), and F1-score (F). The model's performance is evaluated using these measures.

I. Accuracy:
An accurate forecast is one that is proportional to the number of accurate predictions as well as the total number of predictions. True positives (TP) and true negatives (TN) are the components that make up accurate or accurate forecasts. Each and every forecast is made up of the total number of instances, both positive (P) and negative (N). P is made up of true positives (TP) and false positives (FP), whereas N is made up of true negatives (TN) and false negatives (FN). The mathematical representation of this may be described by utilizing the formula that is provided in Eq. 1.

$$A = \frac{True\ Positive + True\ Negative}{Total\ Number\ Of\ Predictions} \quad (1)$$

II. Precision:
The word "precision," which is a measure, refers to the frequency with which a machine learning model correctly predicts the positive class. This is considered to be the definition of the phrase. By dividing the total number of cases that the model predicted as positive (including both true and false positives) by the number of instances in which the model properly anticipated a positive result (true positives), one can evaluate the level of precision that the model has. The mathematical representation of this may be described by utilizing the formula that is provided in Eq. 2.

$$P = \frac{True\ Positive}{True\ Positive\ + Flase\ Positive} \quad (2)$$

III. Recall:
The recall is a quantitative measure that assesses the frequency at which a machine learning model accurately recognizes positive occurrences (true positives) out of all the actual positive samples in the dataset. Recall may be computed by dividing the count of genuine positives by the count of positive cases. The latter includes both true positives (correctly detected cases) and false negatives (cases that were overlooked). The mathematical representation of this may be expressed using the formula given in Eq. 3.

$$R = \frac{True\ Positive}{True\ Positive + Flase\ Negative} \tag{3}$$

VI F1-Score:
It is a measure that is used to evaluate the accuracy and recall of a model in each specific class, and the F1-score is just that. When dealing with datasets that have unequal class distributions, it is often used. We make use of the F1-score as a criterion for evaluation in order to demonstrate the effectiveness of the technique that we have proposed. Equation 4 is used to calculate precision.

$$F1 = 2 \times \frac{Precision \times Recall}{Precision + Recall} \tag{4}$$

4.3 *Performance Evaluation*

This research used a dataset of approximately 349 raw data points split into four classifications to assess the efficacy of five different federated models for ovarian cancer detection. Several criteria are used to assess the performance: in Sect. 4.3.1, performance matrices are analyzed; in Sect. 4.3.2, hyper-parameter tuning is examined; and in Sect. 4.3.3, confusion matrix analysis concludes the discussion. In Sect. 4.3.4 computational complexity is shown.

4.3.1 Performance Matrices

The models that have been assessed include Logistic Regression, XGBoost, SVM, NB, and RF. Of all the models that were assessed, FedXGBoost had the greatest test accuracy at 85.75%, followed by FedSVM at 89.69%, FedRF at 72.08%, FedNB at 91.32%, and Logistic Regression at 95.71%. FedRF did the worst, with an accuracy score of just 72.08%, even if the obtained scores are satisfactory and higher than those of the present study. We used the macro precision metric for the precision measurement, which is determined by averaging the precision scores across all classes. The macro precision determines the accuracy for every class separately before averaging these results without weight. This transaction shows that all models, with the exception of FedRF, performed well; FedRF's values ranged from 72.00% to 95.71% for FedLogisticRegression [Table 3].

4.3.2 Hyper-Parameter Tuning

In machine learning, choosing the best set of hyperparameters for a particular method or model is known as hyper-parameter tuning. A variety of hyperparameters have

Table 3 Federated models' performance matrices on test and train data

Federated model	Test data				Train data			
	Accuracy	Precision	Recall	F1-score	Accuracy	Precision	Recall	F1-score
FedLogisticRegression (%)	98.21	97.24	99.30	98.26	95.71	94.59	97.22	95.89

Table 4 Hyper-parameters of evaluated models

Federated model	Rate of learning	Optimizer	Clients	Beta7	Beta8	Epsilon	Number of rounds
FedLogisticRegression	0.001	FedOPT	3	0.9	0.999	1.00E-08	20

been tweaked to get the best results for the models. The federated model name, learning rate, optimizer, number of clients, beta1, beta2, and epsilon are all shown in Table 4. A number of carefully chosen hyperparameters were used in this research to maximize the performance of federated models. One crucial hyper-parameter that governs how fast the model changes its parameters in response to the calculated oversight gradient is the rate of learning. Since 0.001 has been shown to be an efficient learning rate in several deep learning applications, it was selected. In order to provide an equitable comparison across models, the same number of rounds and clients were used for each model.

4.3.3 Confusion Matrix Analysis

Confusion Matrix Analysis: In machine learning, one of the most important tools for evaluating classification models is the confusion matrix, which is sometimes called the error matrix. It provides a bird's-eye perspective of how a model's predictions stack up against the real data labels. The confusion matrix is an essential tool for assessing the precision and efficiency of the classification model as it displays the several kinds of mistakes produced by the model, like as false negatives, false positives, true negatives, and true positives. Figure 2 illustrates the confusion matrix, providing a visual depiction of the achievements of the FedLogisticRegression model on both the test and training datasets.

4.3.4 Computational Complexity

Training time is an important parameter for evaluating model performance and analyzing computational complexity. When training models using the open-source resources of Google Collaboratory, we carefully considered how many epochs to use

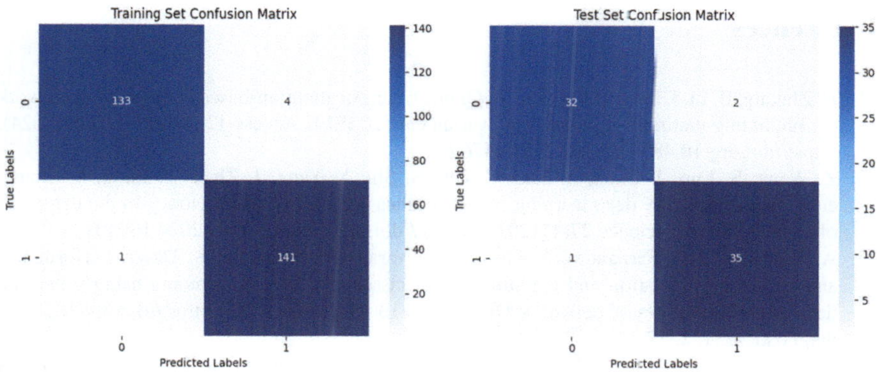

Fig. 2 Confusion Matrix for test and training data

and how to best use the available resources. The whole training time, which includes both CPU and Wall time, was used to guarantee a precise evaluation.

- CPU time: the amount of time a program or process uses up the CPU when running in the mode of user. It sums up the computing effort of the users in its most basic form. Also, it's utilized for running code in privileged modes, such as system calls or kernel mode. 62.5 milliseconds is the amount of time spent on the CPU.
- Wall time: Wall time, also known as real-time, provides an exact measurement of the whole elapsed time, including not only execution time but also system calls, I/O activities, and any possible delays. The wall time is 194 milliseconds.

5 Conclusion and Future Work

This research introduced a federated learning architecture that uses CSV-based data for ovarian cancer detection in a safe and collaborative manner. Our strategy proved to be effective and resilient in detecting ovarian cancer, as we attained an accuracy of 95.71% by using the FedLogisticRegression model. In addition to improving the model's performance and generalizability, our method allows for collaborative learning across several institutions, which further protects patients' privacy by keeping their data localized. To ensure compliance with privacy requirements and to leverage the collective knowledge of participating companies, the federated learning architecture efficiently helps to lessen the dangers of exchanging data.

A restricted amount of data is used, namely, textual data from 349 patients. Increasing the size of the dataset so that it has a wider range of patient demographics and clinical information, which may assist in the development of models that are more generalized and robust.

References

1. H. Zhuang, B. Li, J. Ma, P. Monkam, W. Qian, D. He, An attention-based deep learning network for predicting platinum resistance in ovarian cancer. IEEE Access **12**, 41000–41008 (2024). https://doi.org/10.48550/arXiv.2311.04769
2. Z. Wang, S. Luo, J. Chen, Y. Jiao, C. Cui, S. Shi, Y. Yang, J. Zhao, Y. Jiang, Y. Zhang, et al., Multi-modality deep learning model reaches high prediction accuracy in the diagnosis of ovarian cancer. Iscience **27**(4) (2024). https://doi.org/10.1016/j.isci.2024.109403
3. A. Kodipalli, S.L. Fernandes, V. Gururaj, S. Varada Rameshbabu, S. Dasar, Performance analysis of segmentation and classification of ct-scanned ovarian tumours using u-net and deep convolutional neural networks. Diagnostics **13**(13), 2282 (2023). https://doi.org/10.3390/diagnostics13132282
4. V. Kiruthika, S. Sathiya, M. Ramya, K.S. Sankaran, An intelligent machine learning approach for ovarian detection and classification system using ultrasonogram images. Eng. Sci. **23**, 879 (2023). https://doi.org/10.30919/es8d879
5. S. Lheureux, M. Braunstein, A.M. Oza, Epithelial ovarian cancer: evolution of management in the era of precision medicine. CA: Cancer J. Clin. **69**(4), 280–304 (2019). https://doi.org/10.3322/caac.21559
6. M. Köbel, S.E. Kalloger, S. Lee, M.A. Duggan, L.E. Kelemen, L. Prentice, K.R. Kalli, B.L. Fridley, D.W. Visscher, G.L. Keeney et al., Biomarker-based ovarian carcinoma typing: a histologic investigation in the ovarian tumor tissue analysis consortium. Cancer Epidemiol. Biomark. & Prev. **22**(10), 1677–1686 (2013). https://doi.org/10.1038/bjc.2014.567
7. D. Mistry, M.R. Tonmoy, M.S. Anower, A.T. Hasan, Federated transfer learning for vision-based fall detection, in *International Conference on Big Data, IoT and Machine Learning* (Springer, 2023), pp. 961–975. https://doi.org/10.1007/978-981-99-8937-9_64
8. H.H. Maria, A.M. Jossy, S. Malarvizhi, A hybrid deep learning approach for detection and segmentation of ovarian tumours. Neural Comput. Appl. **35**(21), 15805–15819 (2023). https://doi.org/10.1007/s00521-023-08569-y
9. D. Wang, C. Quek, G.S. Ng, Ovarian cancer diagnosis using a hybrid intelligent system with simple yet convincing rules. Appl. Soft Comput. **20**, 25–39 (2014). https://doi.org/10.1016/j.asoc.2013.12.018
10. M.M. Ahamad, S. Aktar, M.J. Uddin, T. Rahman, S.A. Alyami, S. Al-Ashhab, H.F. Akhdar, A. Azad, M.A. Moni, Early-stage detection of ovarian cancer based on clinical data using machine learning approaches. J. Personal. Med. **12**(8), 1211 (2022). https://doi.org/10.3390/jpm12081211
11. B.L. Wen, M.A. Brewer, O. Nadiarnykh, J. Hocker, V. Singh, T.R. Mackie, P.J. Campagnola, Texture analysis applied to second harmonic generation image data for ovarian cancer classification. J. Biomed. Opt. **19**(9), 096007–096007 (2014). https://doi.org/10.1117/1.jbo.19.9.096007
12. P. Yasodha, N. Ananthanarayanan, Analysing big data to build knowledge based system for early detection of ovarian cancer. Indian J. Sci. Technol. (2015). https://doi.org/10.17485/ijst/2015/v8i14/65745
13. A. Enshaei, C. Robson, R. Edmondson, Artificial intelligence systems as prognostic and predictive tools in ovarian cancer. Ann. Surg. Oncol. **22**, 3970–3975 (2015). https://doi.org/10.1245/s10434-015-4475-6
14. A. BenTaieb, H. Li-Chang, D. Huntsman, G. Hamarneh, A structured latent model for ovarian carcinoma subtyping from histopathology slides. Med. Image Anal. **39**, 194–205 (2017). https://doi.org/10.1016/j.media.2017.04.008
15. M. Wu, C. Yan, H. Liu, Q. Liu, Automatic classification of ovarian cancer types from cytological images using deep convolutional neural networks. Biosci. Reports **38**(3), BSR20180289 (2018). https://doi.org/10.1042/bsr20180289

16. L. Zhang, J. Huang, L. Liu, Retracted article: Improved deep learning network based in combination with cost-sensitive learning for early detection of ovarian cancer in color ultrasound detecting system. J. Med. Syst. **43**(8), 251 (2019). https://doi.org/10.1007/s10916-019-1356-8
17. Q. Chen, S. Wang, J.-H. Lang, Development and validation of nomograms for predicting overall survival and cancer-specific survival in patients with ovarian clear cell carcinoma. J. Ovar. Res. **13**, 1–9 (2020). https://doi.org/10.1186/s13048-020-00727-3
18. M. Akazawa, K. Hashimoto, Artificial intelligence in ovarian cancer diagnosis. Anticancer Res. **40**(8), 4795–4800 (2020)
19. R.M. Ghoniem, A.D. Algarni, B. Refky, A.A. Ewees, Multi-modal evolutionary deep learning model for ovarian cancer diagnosis. Symmetry **13**(4), 643 (2021). https://doi.org/10.21873/anticanres.14482
20. K.R. Kasture, et al., A new deep learning method for automatic ovarian cancer prediction & subtype classification. Turkish J. Comput. Math. Educ. (TURCOMAT) **12**(12), 1233–1242 (2021). https://doi.org/10.17762/turcomat.v12i12.7569
21. N. Taleb, S Mehmood, M. Zubair, I. Naseer, B. Mago, M.U. Nasir, Ovary cancer diagnosing empowered with machine learning, in *2022 International Conference on Business Analytics for Technology and Security (ICBATS)* (IEEE, 2022), pp. 1–6. https://doi.org/10.1109/ICBATS54253.2022.9759010
22. Y. Qiu, M. Tan, S. McMeekin, T. Thai, K. Ding, K. Moore, H. Liu, B. Zheng, Early prediction of clinical benefit of treating ovarian cancer using quantitative ct image feature analysis. Acta Radiologica **57**(9), 1149–1155 (2016). https://doi.org/10.1177/0284185115620947
23. Q. Chen, S. Wang, J.-H. Lang, Development and validation of nomograms for predicting overall survival and cancer-specific survival in patients with ovarian clear cell carcinoma. J. Ovarian Res. **13**, 1–9 (2020). https://doi.org/10.1186/s13048-020-00727-3
24. B.S. Diba, J.D. Plabon, M.M. Rahman, D. Mistry, A.K. Saha, M. Mridha, Explainable federated learning for privacy-preserving bangla sign language detection. Eng. Appl. Artif. Intell. **134**, 108657 (2024). https://doi.org/10.1016/j.engappai.2024.108657
25. D. Mistry, M.F. Mridha, M. Safran, S. Alfarhood, A.K. Saha, D. Che, Privacy-preserving on-screen activity tracking and classification in e-learning using federated learning. IEEE Access (2023). https://doi.org/10.1109/ACCESS.2023.3299331
26. H.B. McMahan, E. Moore, D. Ramage, B.A. y Arcas, Federated learning of deep networks using model averaging (2016). arXiv:abs/1602.05629. https://doi.org/10.48550/arXiv.1602.05629
27. D. Mistry, J.D. Plabon, B.S. Diba, S. Mukta, M.F. Mridha, Federated learning-based architecture for personalized next emoji prediction for social media comments. IEEE Access (2024). https://doi.org/10.1109/ACCESS.2024.3448470
28. M.S.H. Shovon, Z. Sultana, J. Shin, M.A. Hamid, D. Mistry, Incorporating residual connections into a multi-channel cnn for lung cancer detection in digital pathology, in *Data-Driven Clinical Decision-Making Using Deep Learning in Imaging* (Springer Nature Singapore, Singapore, 2024), p. 189. https://doi.org/10.1007/978-981-97-3966-0_10
29. M.G. Kibriya, D. Mistry, D. Mistry, M.T. Rahman, S.H. Ovi, A. Tabassum, S. Hossain, Privacy-preserving vision-based detection of pox diseases using federated learning, in *Data-Driven Clinical Decision-Making Using Deep Learning in Imaging* (Springer Nature Singapore, Singapore, 2024), pp. 105–122. https://doi.org/10.1007/978-981-97-3966-0_6. https://doi.org/10.1007/978-981-97-3966-0_6
30. A. Tabassum, S.H. Ovi, S. Hossain, M.R. Tonmoy, M.S.H. Shovon, M.R. Hussein, D. Mistry, Privacy preserving breast cancer prediction with mammography images using federated learning, in *Data-Driven Clinical Decision-Making Using Deep Learning in Imaging*. (Springer Nature Singapore, Singapore, 2024), pp. 227–245. https://doi.org/10.1007/978-981-97-3966-0_12

Privacy-Preserving Rheumatism Detection Using Federated Learning

Bidita Sarkar Diba, Tasnim Jahin Mowla, Nazneen Nahar, Fahad Ahmed, and Durjoy Mistry

Abstract Rheumatism, a condition affecting joints, muscles, and connective tissues, significantly impacts the quality of life for millions worldwide. Detecting rheumatism accurately and effectively is crucial for timely intervention and treatment. Conventional diagnostic techniques present privacy issues in that most of them involve the consolidation and collection of patient data. However, there is a lack of focus on the detection of rheumatism through investigations that uphold the privacy of patients. This research utilizes Federated Optimization (FedOpt) integrated with Support Vector Machine (SVM) to build an effective FedOpt-SVM classification model that ensures patient privacy. In this regard, the dataset was preprocessed with state-of-the-art techniques such as data normalization and feature selection. The FedOpt-SVM model resolved with a high accuracy of 95.62% for the training set while the precision was 94.44%, the recall was 100%, and the F1-score was 97.14%. As for the test results, the proposed model demonstrated commendable results, achieving an accuracy of 87.50% and precision, recall, and F1-score of 87.50%, 100%, and 93.33%, respectively. Our methodology leverages the strengths of federated learning to preserve data privacy without compromising accuracy. Through extensive experimentation, our model achieved outstanding results, significantly advancing rheumatism detection. This research makes substantial contributions to the fields of medical diagnosis and privacy-aware machine learning, offering a robust framework that benefits both the healthcare community and society as a whole.

B. S. Diba · T. J. Mowla · N. Nahar · D. Mistry (✉)
Department of CSE, University of Asia Pacific, Dhaka, Bangladesh
e-mail: durjoy@uap-bd.edu

B. S. Diba
e-mail: bidita@uap-bd.edu

T. J. Mowla
e-mail: 21101069@uap-bd.edu

N. Nahar
e-mail: 21201145@uap-bd.edu

F. Ahmed
Advanced Machine Intelligence Research Lab, Dhaka, Bangladesh
e-mail: fahadahmed@uap-bd.edu

M. F. Mridha and N. Dey (eds.), *Federated Learning in Health Care Technology*, Studies in Computational Intelligence 1216, https://doi.org/10.1007/978-981-96-8353-6_8

Keywords Federated learning · Rheumatism · Privacy-preservation · FedOpt · Support vector machine

1 Introduction

Diseases affecting the joints, muscles, and connective tissue are collectively referred to as rheumatism. Although many different conditions fall under the umbrella of rheumatism, they are all characterized by joint inflammation and pain [1]. The more accurate phrase "rheumatic disease" is preferred these days. There are two categories of rheumatic disorders: autoimmune and non-autoimmune. Osteoarthritis (OA), fibromyalgia, and gout are instances of non-autoimmune rheumatic disorders that frequently lead to joint pain, stiffness, and inflammation. The term "autoimmune rheumatic diseases" refers to conditions whereby the immune system of the patient attacks its tissues, resulting in chronic inflammation. These diseases include rheumatoid arthritis (RA), ankylosing spondylitis, psoriatic arthritis, systemic lupus erythematosus (SLE), Sjögren's syndrome, and Polymyalgia rheumatica [1]. Mancuso et al. [2] state that patients with rheumatic diseases reported experiencing worsening symptoms as a result of both direct and indirect effects, including psychological stress and worry related to the COVID-19 pandemic, increased fatigue, and deteriorating musculoskeletal function.

Rheumatoid arthritis was estimated to affect 17.6 million people globally in 2020, with a prevalence rate of 208.8 cases per 100,000 individuals, according to a study by Black et al. [3], and by 2050, it is predicted that 31.7 million people will have rheumatoid arthritis worldwide. According to a study by Tian et al. [4], the prevalence of SLE in the newly diagnosed population was 0.40 million per calendar year and 5.14 (1.4 – 15.13) for every 100,000 person-years globally. Grygiel-Górniak et al. [5] claim that rheumatic diseases in women of reproductive age can affect the course of pregnancy and raise risk factors, necessitating the best possible disease control before conception. Bramanti et al. [6] concluded that parents who had children with pediatric rheumatic disorders (PRDs) claimed their kids were more hyperactive and emotionally disturbed than parents who had healthy kids, and that during the COVID-19 pandemic, both groups' emotional symptoms intensified. Xie et al. [7] suggest that rheumatism may cause cardiac hemangioma.

Federated learning reduces the amount of sensitive data exposed to central servers by performing model training locally on user devices. Sophisticated cryptographic techniques, such as client-side encryption, ensure data privacy during local calculations. To preserve model accuracy and protect individual data privacy, controlled noise is introduced into aggregated updates through the use of differential privacy methods. Explicit consent systems preserve user autonomy, permitting engagement in and withdrawal from federated learning tasks in accordance with strict privacy laws and guaranteeing strong secrecy in decentralized learning environments [8].

This paper provides an innovative approach to rheumatism detection that prioritizes user privacy protection by exploiting the highly sophisticated features of federated learning. Our work differs from previous studies in this field in that it strongly commits to the safety of user data, filling a significant gap in the literature. This methodology enhances the accuracy and reliability of disease detection while simultaneously providing a guarantee of user privacy protection at every step of the process. The benefits of federated learning extend beyond privacy protection. Since the data does not need to be centralized for analysis, and the findings from various sources can be combined into a single aggregate, it provides better reliability and versatility of the detection model. This work marks a paradigm for the growth of the relevant subsector and is considered as a stepping stone toward constructing a safer, more dependable, and privacy-sensitive rheumatism detection process. Incorporating higher levels of privacy preservation and analytical approaches, our method increases user trust while also improving the detection of rheumatism in terms of effectiveness and efficiency.

We will now sum up the various arguments made in the paper and give a summary of what the paper adds:

- This paper is aimed at implementing FedOpt together with SVM, called FedOpt-SVM, to classify more accurately the affection status of Rheumatoid Arthritis and privacy protection.
- Our methodology will incorporate Federated Optimization to enhance the performance of the SVM model on the federated dataset without any loss of privacy or security.
- Experimental results show that our approach achieves robust performance, which validates its efficacy in many real-world scenarios where privacy is very important in the data.
- This work advances federated learning in medical applications, particularly in the area of rheumatoid arthritis classification, by developing a practical solution able to strike a balance between privacy requirements and model accuracy.

The rest of the paper is structured as follows: Related Work explores methods and innovations for diagnosing and managing rheumatism using deep learning and machine learning. Conventional Privacy-Preserving Methods address securing sensitive dataset information related to patient health. Methodology provides a dataset description, preprocessing steps, the proposed framework, and the application of the Fed algorithm. Result & Analysis presents experimental results, evaluation metrics, and comparisons with previous research. Conclusion and Future Work highlights findings, contributions, and potential directions for further investigation.

2 Literature Review

Rheumatism is a complex condition with a significant impact on patients' quality of life, and its diagnosis and management often pose challenges. However, advancements in deep learning (DL) and machine learning (ML) offer remarkable possibilities for enhancing rheumatism diagnosis, prognosis, and personalized treatment. This section explores a diverse array of methodologies, machine learning, deep learning, and their integration.

2.1 Advancements in Biomarker Identification

Making use of machine learning and biomarkers, Bianchi et al. [9] concentrated on making an early diagnosis of temporomandibular joint osteoarthritis (TMJ OA). Data science techniques were used to examine 52 clinical, biological, and radiomic indicators using machine learning techniques such as Logistic Regression, Random Forest, LightGBM, and XGBoost. An accuracy of 0.823 was obtained using XGBoost in addition to the LightGBM model.

Luque-Tévar et al. [10] took advantage of machine learning algorithms to forecast TNFi responsiveness in RA patients. A total of 104 patients as well as 29 donors in good health were enrolled. In accordance with the research, patients with RA who are on anti-TNF medication group together according to changes in their molecular profiles and how effectively their treatments work. Machine learning algorithms accurately predicted TNFi response, with an increase in accuracy when combined into a mixed model. The model identified signatures involving oxidative stress markers, NETosis products, inflammatory cytokines, and microRNAs as TNF response predictors. The study's weaknesses include the following: the involvement of other circulating macromolecules in treatment response cannot be ruled out because a comprehensive blood profile encompassing many biomarkers was not conducted. Due to the restricted amount of samples and clinical variability of RA patients, the study's conclusions require validation in larger cohorts. Additional investigations are necessary to determine the cellular origins of these molecules' changed expression and to comprehend the mechanisms underlying their altered expression following TNFi therapy.

According to Wen et al. [11], ZDHHC18 and IL2RB may be used as blood biomarkers for the early detection of ankylosing spondylitis (AS). Commonly differentially expressed genes in AS patients were screened using machine learning algorithms, leading to the establishment of a diagnostic model with C-indexes of at least 0.84. The model's values of the area under the curve (AUC) were 0.86, 0.84, 0.85, and 0.89 in GSE73754, GSE25101, GSE18781, and GSE11886, in that order.

Liu et al. [12] addressed the diagnostic limitations of rheumatoid arthritis (RA) by investigating platelet-related mechanisms and identifying potential biomarkers. They employed Weighted Correlation Network Analysis (WGCNA) for investigating expression components within differentially expressed genes (DEGs) using microarray data sources (GSE93272 along with GSE17755) obtained from the GEO database. Among the DEGs examined in Module 2, a noteworthy correlation with platelets was found. The research identified platelet-related signals using KEGG, GO, alongside GSEA enrichment assessment. It then utilized the LASSO technique to create a diagnosis model and evaluated its efficacy using GSE17755 to serve as the validation cohort. Within the cohorts for training and validation, the resulting PRS model showed impressive diagnostic preciseness, with AUC values of 0.801 and 0.979, respectively. This approach helps to improve the understanding of RA and its associated diseases. A few of the study's shortcomings are its limited number of samples and the absence of confirmation from animal and cell research. The PRS model's potential for diagnosis is shown; however, these trials don't provide any support for it. The possible confounding impacts of variables were similarly overlooked in this investigation.

Liu et al. [13] explored the development of biomarkers for rheumatoid arthritis using gene expression datasets from GEO and TCGA databases. Potential biomarkers for RA diagnosis and treatment were found in ITGB2, MMP13, CRTAM, and PTTG1IP by four machine learning algorithms: SVM, XGBoost, RF, and LASSO. With an AUC of 0.894 in training cohorts and 0.843 in testing cohorts, these signatures demonstrated great diagnostic accuracy and distinguished between two different patient groups. CRTAM's knockdown suppressed proliferation and invasion in RA cells, indicating its therapeutic potential. CRTAM outperformed classic biomarkers like MMP3, S100A8, and S100A9. Its role in immune dysfunction suggests its involvement in cancer progression, especially among RA patients. The work has constraints due to its dependence on bioinformatics and the need for larger prospective studies to assess the sensitivity and specificity associated with the four new biomarkers. The conclusions drawn may not apply to varying ethnic populations because the bulk of the datasets under analysis are from Western nations. Additionally, the study only evaluated putative biological roles through association analysis, and more trials are needed for validation.

2.2 Advancements in Predictive Modeling and Diagnosis

Vanier et al. [14] set out to create a matrix that could forecast rapid radiographic progression (RRP) among patients with initial rheumatoid arthritis in order to help with treatment decisions. The logistic regression model was selected through a 10-fold stratified cross-validating process in order to maximize the AUC. RRP was present in 20.6% of the 1306 individuals with active premature rheumatoid arthritis who had not been on DMARDs. Rheumatoid factor positivity, RA erosion on X-rays, elevated CRP levels, and swollen joints were found to be the four main predictors. The resulting matrix had a 95% confidence interval that stretched from 0.02 at the lowest to 0.08 at the highest, improving precision and reliability. Furthermore, a p-value of 0.79 indicates that the model calibration was excellent, demonstrating its reliability in predicting RRP in patients with early RA. The research is subject to certain constraints, such as heterogeneity in treatment courses, low positive predictive value, and moderate discriminatory power. Classification mistakes may arise through the discriminating capacity of the final model, which was damaged by missing data, particularly for the outcome (RRP). Imputation strategies for missing data were not used in this investigation, assuming that non-informative absent information had not been met.

Kalweit et al. [15] evaluated AdaptiveNet's performance in forecasting the progression of the disease in patients suffering from rheumatoid arthritis using data from over 9500 patients and 65,000 visits. At an overall accuracy level of 75.6% (SD + −0.7%), AdaptiveNet correctly predicted active disease, which is referred to as DAS28-BSR >2.6. Its sensitivity and specificity were 84.2% (SD + −1.6%) and 61.5% (SD + −3.6%), respectively. Individual DAS28-BSR values were forecasted by regression having a mean squared error (MSE) of 0.9(SD + −0.05), signifying an 8% discrepancy between the estimated and actual values. When compared to SVM, random forest, and linear regression, AdaptiveNet performed more than 7% better in MSE. The lack of comparison with statistical models that do not use machine learning, the need for greater specificity, the difficulties in handling incomplete and heterogeneous data for machine learning, the possibility of overfitting in classification tasks, the limited predictive power of medication, and the uncertainty surrounding DL's drug response prediction are some of the drawbacks.

Matsuo et al. [16] used information from blood tests and ultrasound (US) exams to predict relapse in patients with RA. Based on disease activity, they classified 210 RA patients who were in remission at baseline and followed up with them for two years (n = 150) and the relapse (n = 60) groups. Three machine learning classifiers (ML) were used: logistic regression, random forest, and XGBoost. 73 features, such as US examination data, blood test results, and patient information, were used in the algorithms. With an AUC of 0.747, XGBoost outperformed Logistic Regression and Random Forest, which had AUCs of 0.719 and 0.701, respectively.

Queiro et al. [17] sought to forecast patients with recent onset of psoriatic arthritis (PsA) with moderate-to-high disease activity. It used 158 patients over two years and three machine learning algorithms: XGBoost, random forest, and logistic regression. PsAID score, tender joint count, sex, and physical activity were important predictors. After a follow-up, 21.2% of patients had moderate-to-severe disease, up from 20.8%. High validity was demonstrated by machine learning algorithms, especially in terms of sensitivity (98%; 95% CI: 86.89–100.00). The most significant predictor, PsAID, surfaced, indicating its usefulness in clinical practice. The study also suggested specific approaches to meet the needs of female PsA patients.

Batool et al. [18] presented an automated tool for diagnosing Rheumatoid Arthritis (RA) and assessing its disease activity levels using machine learning classifiers. Clinical records, lab results, and demographic information from 104 RA patients were gathered from a nearby Multan hospital. They compared AdaBoost, SVM, Random Forest, Naïve Bayes, and Decision Trees—five supervised machine learning classifiers. The purpose of this research was to determine the best algorithm for early RA diagnosis. Compared to other algorithms, AdaBoost performed better, achieving 87.1% accuracy and 88% precision in classifying RA patients and non-RA patients, 93% accuracy and 82% precision in classifying high and moderately active diseases, and 63% accuracy and 100% precision in classifying low active diseases. Some of the drawbacks of this study are the limited number of samples and inadequate data for some variables. Notwithstanding the use of machine learning and artificial intelligence models, overfitting is a possibility, particularly with XGBoost since it may have been overly customized to the training set. The discrepancy between the training and test subsets may have an impact on the reported performance disparities between the random forest and XGBoost models, presumably as a result of the dataset's small sample size.

2.3 Enhanced Imaging Analysis Techniques

To identify structural alterations suggestive of hand osteoarthritis in radiography pictures, Üreten et al. [19] developed an automated technique founded on transfer-learning-based convolutional neural networks. Pretrained AlexNet, GoogLeNet, and VGG-19 networks were used to create multilayer convolutional neural network models centered on a transfer learning technique. The models obtained accuracies of 93.2% (AlexNet), 94.3% (GoogLeNet), and 96.6% (VGG-19) after being trained on a dataset of 332 images and validated with 88 images. For AlexNet, GoogLeNet, and VGG-19, the corresponding sensitivities were 0.9167, 0.9184, and 0.9574, and the corresponding specificity values were 0.9500, 0.9744, and 0.9756. Because the dataset used in this work ranged from early to late OA, and since the authors had not coded their CAD approach to attempt an assessment of the severity of OA employing the Kellgren-Lawrence scoring system, they could not effectively discuss the reliability of the methodology for the diagnosis of early OA.

Using hand radiography images, Mate et al. [20] developed a Convolutional Neural Network (CNN) that can identify the stage of Rheumatoid Arthritis (RA). With 290 radiography image datasets, the CNN's accuracy was 94.46%. The network also showed a sensitivity of 0.95 and specificity of 0.82, demonstrating the ability to classify RA phases in hand X-rays correctly.

After Üreten et al. [21] used the You Only Look Once (YOLO) algorithm, a pre-trained VGG-16 network was used for transfer learning to extract hand images from the initial radiographs without erasing any data. In comparison to normal hand radiographs, the classification results for rheumatoid arthritis showed 90.7% accuracy, 92.6% sensitivity, 88.7% specificity, 89.3% precision, and a 0.97 AUC. The results showed 90.8% accuracy, 91.4% sensitivity, 90.2% specificity, 91.4% precision, and a 0.96 AUC for osteoarthritis compared to normal hand radiographs. An accuracy of 80.6% was attained by combining osteoarthritis, rheumatoid arthritis, and normal hand radiographs. The study's weak point is the lack of determination of the modified Sharp score and disease activity scores for OA patients based on their radiographs. The patients' radiographs showed early and late radiological alterations from RA, but the authors neglected to compute the modified Sharp score, which is one of the study's weak points. Similarly, they did not use the Kellgren and Lawrence scoring system to determine disease activity scores in OA patients.

With the use of digital X-ray images, Srinivasan et al. [22] sought to evaluate the efficacy of an AI-based framework for knee RA detection and grading. 3172 photos from the BioGPS database were acquired, and participants over 50 with RA symptoms were included. ResNet-101 extracted features, Faster-CRNN architecture found joint space narrowing areas, and VGG-16 graded the severity of RA. Medical professionals used a consensus-based decision score to grade the images. The model achieved was able to classify the total knee RA intensity with 99.10% accuracy, also obtaining a sensitivity of 97.3%, specificity of 98.2%, precision of 98.1%, and dice score of 90.1%. It also successfully identified the marginal knee JSN region by achieving 98.97% accuracy.

Using deep learning models, Shuai et al. [23] used Fourier transform infrared (FTIR) spectroscopy on dried serum samples to identify rheumatoid arthritis and ankylosing spondylitis at an early stage. To efficiently fuse local features, three multi-scale convolutional modules based on the multi-scale convolutional neural network (MSCNN) were created. 243 serum samples were gathered for the study, and four classification models were created. With an accuracy of 0.93 and an AUC of 0.99, the MSCNN model proved to have better diagnostic performance. This method demonstrated potential for quick and non-invasive diagnosis of rheumatism; however, the study's limitations, including its small sample size, were acknowledged.

2.4 Understanding COVID-19 Impacts on Rheumatic Patients

Ye et al. [24] examined data from 2326 COVID-19 patients, 21 of whom also had rheumatoid arthritis at the same time. In rheumatic cases, respiratory failure was more frequent (38% vs. 10%), but the length of hospital stay and death rates between the rheumatic and non-rheumatic groups were similar, according to the results. Fever (76%) and exhaustion (43%) were common symptoms, as was diarrhea (23%). While in the hospital, four rheumatoid arthritis patients had flare-ups in their condition. Only 5% of patients with rheumatic cases, showed leucopenia, whereas lymphocytopenia was observed in 57% of rheumatic patients. Before COVID-19 was identified, five rheumatic cases took hydroxychloroquine; none of them developed a serious illness. The limited number of samples and the fact that none of the 21 individuals in this study had long-term COVID-19 treatment provide limitations. Additionally, because there are so few instances with a clear history of CTD-ILD, the risk of COVID-19 lung injury in rheumatoid arthritis patients with pre-existing ILD was not evaluated. Owing to the small number of patients in some disease groups, the study also included a collective of rheumatic diseases that were not further subcategorized. As a result, it is unable to offer details regarding how these drugs affect COVID-19 susceptibility and clinical advancement.

Tripathy et al. [25] investigated the possible link between rheumatoid arthritis and SARS-CoV-2 infection. To investigate the relationship between SARS-CoV-2 infection and RA, a non-systematic review was conducted on the material found in the Google Scholar and PubMed databases. The study found a possible correlation between rheumatoid arthritis (RA) and elevated angiotensin II levels, which in turn led to the overregulation of pro-inflammatory cytokines that include TNF-α, IL-6, and IL-8. COVID-19 infections may be made more difficult by variables such as increased ROS production, lymphocytic activation, and immune system suppression in RA patients taking DMARDs. JAK/STAT, NF-κB, and MAPK pathway activation amid SARS-CoV infection may cause genetic factors linked to RA or exacerbate immune responses. Research indicates a possible co-occurrence of severe RA and COVID-19, particularly in genetically predisposed individuals, and points to a predisposition for severe RA in COVID-19 victims.

Burns et al. [26] utilized machine learning to determine indicators of risk for severe COVID-19 among individuals with rheumatoid arthritis (RA). The XGBoost, logistic regression, random forest, and SVM models were fitted using five-fold cross validation. 23.1% of 4,295 patients with COVID-19 who had rheumatoid arthritis were found to have severe cases. Notably, compared to patients with mild cases, those with severe COVID-19 showed a higher Charlson Comorbidity Index (CCI) (3.8 [SD 2.4] vs. 2.4 [SD 1.8], $p < 0.001$) and average age (71.6 [SD 10.3] vs. 63.4 [SD 13.7] years, $p < 0.001$). Moreover, there were more males in severe cases than in mild ones (29.1% vs. 18.5%, $p < 0.001$). The best-performing model (XGBoost, AUC = 75.64) yielded the top 15 predictors. In the setting of machine learning, the cohort that satisfies their eligibility requirements is a rather tiny sample. Furthermore, because of the nature of the data, diagnosis definitions are dependent only on ICD-10-CM codes, and unmeasured factors (such as lab results and vital signs) may exist. These are the study's limitations.

According to research done in Dhaka by FERDOUS et al. [27] between June 13 and October 12, 2020, 7.78% of patients with rheumatoid arthritis had COVID-19 infection. At 42.15±13.04 years, the mean age of the group was male. In 69.23% of cases, RT-PCR verified the presence of COVID-19, with the remaining cases being diagnosed through clinical features and chest imaging. For patients who tested positive for COVID-19, fever was the most common presentation. Spondyloarthritis (38.56%) and knee osteoarthritis (23.88%) were the most prevalent rheumatic diseases, respectively, followed by psoriatic arthritis (15.38%). Among COVID-19 cases, comorbid conditions such as diabetes (38.56%), hypertension (34.46%), and bronchial asthma (33.88%) were common. To treat rheumatic diseases, non-specific medications, and continuous DMARDs or biologics were used. Nevertheless, there are some limitations, such as incomplete medication data, potential selection bias, and the exclusion of some COVID-19-positive rheumatic patients. However, there are certain restrictions. The study did not cover all rheumatoid arthritis patients with a COVID-19-positive status, and selection bias may have crept in during the consecutive sampling of subjects. Other than the medications provided during the study period, no particular information was available regarding the other drugs taken by the patients or how long they took them.

Wang et al. [28] examined the dynamics of neutralizing antibodies and SARS-CoV-2 IgG antibodies in patients suffering from RMD following COVID-19 infection, suggesting possible variations in immune responses. The study included 104 HC and 292 RMD patients in total. RMD patients showed significantly lower levels of anti-S1/S2 IgG and NAb at 60 and 90 days after COVID-19 infection ($p<0.001$) compared to the HC group. Anti-S1/S2 IgG antibody levels held steady, but in RMD patients, there was significant variation in NAb levels between the 60th and 90th days. Regardless of the particular RMD subtypes, a logistic regression analysis showed that prior administration of b/tsDMARDs, immunosuppressants, and glucocorticoids (GC) stood out as independent risk factors associated with decreased anti-S1/S2 IgG and NAb levels. There are various restrictions on the study. Because it depends on observational data, a clear cause-and-effect relationship cannot be established. The small sample size hinders the investigation of medication effects across different disorders. Understanding is further hampered by the lack of information on the severity of the COVID-19 infection and the antibody response 30 days after infection. The study's conclusions are additionally constrained by the fact that it does not contrast antibody levels with other immune responses. These restrictions show how much more research is required.

A brief overview of the literature review is provided in Table 1.

3 Privacy-Preserving Methodologies

Privacy-preserving methodologies in text analysis are crucial for protecting confidential data during the manipulation, examination, or dissemination of textual information. The utilization of privacy-preserving approaches is imperative to shield sensitive data from unauthorized exposure or release. Through the application of these methodologies, entities can reduce the vulnerabilities linked to data breaches, thereby safeguarding both their own interests and the well-being of their clientele from potential financial and reputational repercussions.

When it comes to Homomorphic Encryption, text embedding services let consumers access original content without disclosing it, but there's a chance that information might be compromised as a result of inversion assaults, which could somewhat jeopardize privacy. Comprehensive tests proved that the suggested approach preserves model performance without inflicting any damage. The vector-vector and vector-matrix multiplication computation durations clearly demonstrate the tremendous computational cost associated with homomorphic encryption processes [29]. The existing homomorphic approaches need a lot of user interaction to provide rating results since they are too computationally and communicatively demanding to be efficient [30].

Considering the Privacy-Preserving Record Linkage (PPRL) scenario, tested parameter configurations hold up well with various dataset types. The techniques are designed to protect privacy while preserving the largest possible amount of information. When compared to approaches based on Bloom filters or unencrypted linkage, the accuracy is marginally lower. The approaches include compromises but are nevertheless beneficial and feasible. While the methods are intended to protect privacy, there are concerns about prospective points of attack and absent identifiers that may

Table 1 Summary of literature review

Paper	Year	Dataset	Model	Performance	Limitation
Bianchi et al. [9]	2020	52 markers	LR, RF, LightGBM, XGBoost	XGBoost + LightGBM model's accuracy is 0.823	N/A
Luque-Tévar et al. [10]	2021	104 patients and 29 healthy donors	N/A	N/A	Insufficient blood profile analysis
Wen et al. [11]	2022	N/A	LASSO, LR, SVM-RFE, RF	AUC were 0.86 (GSE73754), 0.84 (GSE25101), 0.85 (GSE18781), 0.89 (GSE11886)	N/A
Liu et al. [12]	2023	GEO	LASSO	AUC of 0.979	Potential confounding effects of variables are not considered
Liu et al. [13]	2024	GEO, TCGA	LASSO, RF, XGBoost, and SVM	AUC = 0.843	Reliance on bioinformatics
Vanier et al. [14]	2020	1306 patients	Logistic Regression	95% confidence interval	Heterogeneity in treatment courses, low predictive value, moderate discriminatory power
Kalweit et al. [15]	2021	9500 patients and 65,000 visits	AdaptiveNet	An overall accuracy of 75.6%	Overfitting, limited predictive power, and uncertainty in drug response prediction
Matsuo et al. [16]	2022	210 RA patients	Logistic regression, XGBoost, and random forest	XGBoost performed best with an AUC of 0.747	N/A
Queiro et al. [17]	2023	158 patients	XGBoost, RF, LR	A sensitivity of (98%; 95% CI: 86.89–100.00)	N/A
Batool et al. [18]	2024	104 RA patients	AdaBoost, SVM, RF, Naïve Bayes, Decision Trees	AdaBoost performed the best with 87.1% accuracy and 88% precision.	Limited sample size and absent data for certain factors

(continued)

Table 1 (continued)

Paper	Year	Dataset	Model	Performance	Limitation
Üreten et al. [19]	2020	After training with 332 images, 88 images were used for validation	Pretrained AlexNet, GoogLeNet, and VGG-19	AlexNet achieved an accuracy of 93.2%, GoogleNet achieved 94.3%, and VGG-19 achieved 96.6%	The lack of coding for their CAD method using the Kellgren-Lawrence scoring system
Mate et al. [20]	2021	290 radiography image datasets	CNN	Accuracy of 94.46%, sensitivity of 0.95 and specificity of 0.82	N/A
Üreten et al. [21]	2022	N/A	YOLO, VGG-16	Accuracy of 90.8% for osteoarthritis	Absence of determining the modified Sharp score and disease activity scores for OA patients based on their radiographs
Srinivasan et al. [22]	2023	BioGPS database	ResNet-101, Faster-CRNN, VGG-16	Dice score of 90.1%, specificity of 98.2%, accuracy of 99.10%, sensitivity of 97.3%, and precision of 98.1%	N/A
Shuai et al. [23]	2024	243 serum samples	Three multi-scale convolutional modules based on MSCNN and four classification models	An accuracy of 0.93 and an AUC of 0.99	Limited sample size
Ye et al. [24]	2020	2326 COVID-19 patients	N/A	N/A	A small sample size
Tripathy et al. [25]	2021	PubMed and Google Scholar databases	N/A	N/A	
Burns et al. [26]	2022	4,295 patients with COVID-19	XGBoost, LR, RF, SVM	XGBoost performed best (AUC = 75.64)	Diagnosis definitions are dependent only on ICD-10-CM codes
FERDOUS et al. [27]	2023	Dhaka, conducted between June 13 and October 12, 2020	N/A	N/A	Incomplete medication data, potential selection bias
Wang et al. [28]	2024	396 participants	LR	N/A	A small sample size, not comparing antibody levels with other immune responses

undermine the safeguarding of privacy. Large-scale links may find the discussed approaches less feasible due to their comparatively poor speed when compared to unencrypted linking [31].

In the course of analyzing Synthetic Text Generation, the application of differential privacy techniques with a robust privacy assurance (privacy budget, $\epsilon = 2$) demonstrates a notable endeavor to safeguard privacy. The synthetic text generation approaches, particularly DistilDP, surpass other baseline methods. The objective of synthetic text generation is to distill insights from the teacher model, indicating a strive to retain maximum information from the original dataset. The utilization of extensive language models such as GPT2-Large highlights the necessity for substantial computational resources. In view of the fact that the synthetic text is intended to reproduce the original dataset, there should be no significant communication complexity as long as the downstream applications are aligned with the format and structure of the generated text [32].

The data anonymization protocol involves moderate information loss due to generalization and suppression inherent in clustering-based k-anonymity. Accuracy is similarly impacted by the introduction of counterfeit-sensitive attributes and necessary generalizations. Data utility is maintained at a moderate level, balancing anonymity and usability, although some utility is inevitably sacrificed. Privacy preservation is high, with robust measures against both internal and external identity disclosure and effective leader collision mitigation. However, the protocol incurs high computational complexity, particularly with larger datasets and multiple sensitive attributes, due to the intensive processes of clustering-based k-anonymity and CSA elimination. Communication complexity is medium to high, given the multiple stages of initialization, leader election, and data collection requiring significant coordination among data owners and group leaders [33].

Federated learning stands out as a robust approach to machine learning that prioritizes privacy and efficiency. It achieves very low information loss by allowing machine learning models to train on decentralized data sources, ensuring that sensitive data remains on individual devices or servers rather than being centralized. This decentralized approach not only preserves user privacy but also facilitates high accuracy by harnessing diverse data sources without compromising data security. In terms of utility, federated learning excels in detecting and managing user activities on computers in a user-friendly manner, aiming to minimize disruptions from unproductive behaviors. The system's architecture ensures very high levels of privacy protection, keeping user data private and secure throughout the training process. Federated learning minimizes computational complexity by leveraging local device resources for model training. Similarly, communication complexity is kept low through optimized protocols that facilitate the transmission of model updates without the need to exchange raw data, further enhancing overall efficiency and security in federated learning implementations [34].

A comparison of different privacy-preserving methodologies is presented in Table 2.

Table 2 Comparison of privacy-preserving methodologies

Techniques	Information loss	Accuracy	Utility	Privacy preserved	Computational complexity	Communication complexity
Homomorphic encryption	Average	Elevated	Elevated	Average	Elevated	Elevated
Privacy-Preserving Record Linkage (PPRL)	Limited	Limited	Elevated	Average	Elevated	Average
Synthetic text generation	Limited	Average	Elevated	Elevated	Average	Limited
Data anonymization	Medium	Average	Average	Elevated	Elevated	Elevated
Federated learning	Limited	Elevated	Average	Elevated	Limited	Limited

4 Methodology

4.1 Dataset Description

The dataset from Boo et al. [35] was used in this investigation, encompassing data from 200 patients diagnosed with rheumatoid arthritis. A university medical hospital with over 1,100 beds in South Korea conducted a study with RA patients as its focus. Individuals who were free of cardiovascular disease (CVD) and had received treatment for any clinical stage of RA were eligible to take part in the trial. In December 2015, 200 RA patients completed questionnaires as part of the study. The majority of the study's subjects were female, with an average age of 52.6. They experienced RA for roughly 6.8 years after receiving a diagnosis at the age of 46.8. Out of 10, their physical functioning score was 0.95. Among the subjects, physical inactivity accounted for 77% of the risk factors for CVD, followed by overweight or obesity (43%), hypertension (39%), and diabetes (7%). Some were receiving therapy to decrease their cholesterol.

4.2 Data Preprocessing

Data preprocessing is a crucial first step in preparing the dataset for machine learning algorithms. This subsection outlines the procedures for preprocessing the dataset in preparation for model training.

4.2.1 Data Normalization

A machine learning preprocessing approach called normalization or scaling involves bringing the range of numerical data to a common scale, usually centered around zero with a standard deviation of 1. In this process, each characteristic needs to be adjusted to have a mean of 0 and a standard deviation of 1. We used "StandardScaler" to normalize the dataset. The formula in this regard is given by Eq. 1:

$$X_{\text{scaled}} = \frac{X - \mu}{\sigma} \tag{1}$$

where

- X: The original dataset
- μ: The mean of the dataset
- σ: The standard deviation of the dataset
- X_{scaled}: The normalized dataset.

4.2.2 Feature Engineering (Polynomial Features)

To capture intricate interactions between features, Polynomial Features were used to engineer polynomial features up to the second degree. This can increase the capacity of the model in learning nonlinear patterns since it creates interaction between variables, thus increasing the feature space. The total number of output features generated by PolynomialFeatures for η input features and degree δ is calculated using the binomial coefficient as shown in Eq. 2:

$$\text{Total number of output features} = \binom{\eta + \delta}{\delta} = \frac{(\eta + \delta)!}{\eta! \cdot \delta!}. \tag{2}$$

4.3 Machine Learning Model

4.3.1 Support Vector Machine

Support Vector Machine was developed to construct the optimality criterion for creating a hyperplane with maximum separation between different classes. It can be generalized to handle nonlinearly separable data by mapping data into a higher-dimensional space employing kernel functions. It has proved to be one of the most robust and efficient algorithms for a wide array of classification tasks [36].

An SVM may be explained by a decision function which could be expressed as Eq. 3:

$$\hat{f} = \text{sign}\left(\sum_{j=1}^{M} \beta_j t_j g(z_j, \mathbf{z}) + d\right) \tag{3}$$

Here, $\beta_j \in [0, D]$ are Lagrange multipliers, positive for support vectors and zero for other training instances. The parameter d stands for the bias term and $g(z_j, \mathbf{z})$ is a kernel function. An instance of such a kernel function is the Gaussian or Radial Basis Function (RBF) as shown in Eq. 4:

$$g(z_j, \mathbf{z}) = \exp\left(-\frac{\|z_j - \mathbf{z}\|^2}{2\tau^2}\right) \tag{4}$$

The parameter D, is known as the box constraint or the penalty, or the regularization term for the error. It is the key when this algorithm involves SVM, since it will strike a balance between maximizing the margin and minimizing the classification errors on the training data. For large values of D, it will have small margins and less misclassification, which may lead to overfitting. Another important hyperparameter is τ in the Gaussian kernel, which decides the shape of the decision boundary. A smaller τ results in a more complex boundary and may also lead to overfitting.

4.4 Framework: Federated Learning

Federated learning is a technique in machine learning whereby a model is trained across many decentralized gadgets, such as cellphones, edge servers, and Internet of Things devices, that might keep local data samples without having to transfer data. Model updates or aggregated gradients are exchanged, not raw data, which could cause privacy issues or use a lot of bandwidth when sent to a central server. Federated learning architecture is illustrated in Fig. 1.

Federated learning executes algorithms created for Federated Learning and systems using the Flower framework, a software library. With federated learning, data is not centralized in one place but rather is trained on a variety of data sources, including edge servers and mobile devices. By sharing model updates or gradients, models can be trained cooperatively without sending data to a centralized server. Federated Learning preserves privacy by securing data security and confidentiality at the source, enabling efficient model training without compromising data privacy [37–41].

With federated learning, a network of decentralized data sources is used to train a shared model in a cooperative setting. To start the learning process, a copy of the global model is transmitted to the devices. Using local datasets, each model is trained independently, and the model weights are then returned to the server. The global server aggregates the weights from various centers and then sends the revised global model back to the centers for additional training. Until convergence, this

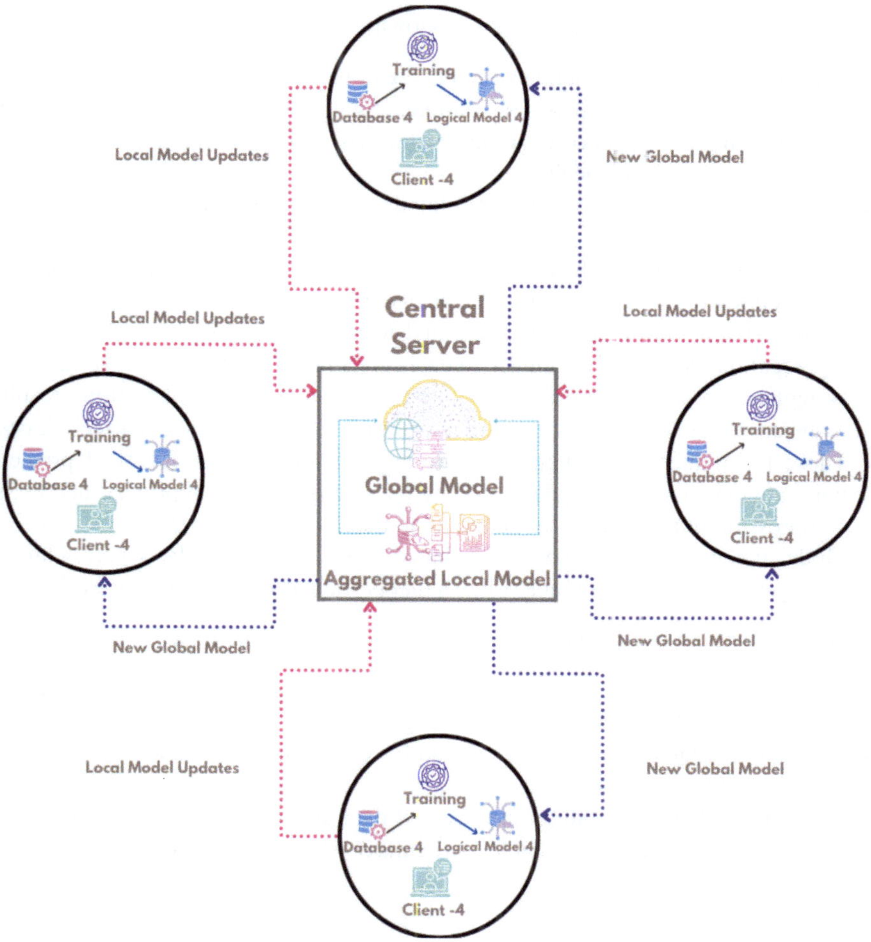

Fig. 1 Federated learning outline

iterative process is carried out. The validation sets are used to evaluate the local model locally. Test data is used to evaluate the global model, and each round's performance indicators are published [42]. Equation 5 is the objective function for federated learning:

$$g(z_1, \ldots, z_N) = \frac{1}{N}\sum_{\ell=1}^{N} g_\ell(z_\ell) \tag{5}$$

Here,

* N: Number of nodes.
* z_ℓ: The model's weights as seen by node ℓ.
* g_ℓ: The local objective function of node ℓ.

A Typical Federated Training Process is shown below [43]:

- Selection of clients: A sample of clients who meet the requirements is selected by the server. Mobile phones, for example, might only connect to the server when they are plugged in, linked to an unmetered wifi connection, and idle in order to prevent harming the device's user.
- Broadcast: The server provides a training program and the current model weights to the chosen clients.
- Client computation: Each selected device computes a local update to the model by executing the training algorithm, which could, for instance, perform SGD over the data that is local (such as in Federated Averaging).
- Aggregation: A total of the device updates is gathered by the server. At this step, stragglers can simply be dropped if there are enough devices submitting results to ensure efficiency. This is also the integration point for many other techniques. Those techniques could include secure aggregation for high levels of privacy, noise augmentation, and update clipping for various degrees of privacy, and lossy compression of aggregates for better communication efficiency.
- Model update: It then updates the shared model locally at the server with an aggregate computed from those clients that participated in the present cycle.

Federated learning is especially helpful in the following situations:

- Privacy is Crucial: Since data stays on users' devices, there are fewer privacy concerns related to centralized data storage.
- Data is Distributed: Data cannot be readily compiled in one place since it is dispersed among several sources.
- Limited Bandwidth: Sending unprocessed data takes a lot of bandwidth, which can be expensive or impracticable, particularly when dealing with big datasets.

4.5 Algorithm: FedOPT

The proposed FedOPT algorithm is a further development of Federated Learning (FL) that was used in this work as a basis, considering the problems of transmission inefficiency and the preservation of personal information privacy. FedOPT improves the defense and efficiency of FL by integrating homomorphic encryption, differential privacy, and efficient encoding algorithms. The FedOPT approach is clarified in the following phases [44]:

- **Initialization:** The client and server optimizers should be created right next to the global model parameters.
- **Sample of Clients:** From the above category of clients, select a random sample of the clientele for the current session of training.
- **Local Training:** The chosen client dictates their data, the local model is synchronized, and the gradients are computed for local updates.

- **Gradient Compression:** To decrease the communication cost, implement the Sparse Compression Algorithm (SCA) to compress gradients.
- **Encryption:** On account of safe aggregation, encode the compressed gradients using lightweight homomorphic encryption.
- **Aggregation:** In order to determine the global gradient, a server combines the encrypted gradients.
- **Decryption:** By employing private keys, the clients decode the global gradient.
- **Update on the Global Model:** The aggregated gradients are utilized via the server for revising the global model.
- **Iteration:** Continue to perform these until integration is achieved.

FedOPT Algorithm: The following pseudocode explains the FedOPT algorithm [45]:

Algorithm 1 FedOPT Algorithm

Enter Data: x_0, SERVEROPT, CLIENTOPT
for each iteration $t = 0, ..., T - 1$ **do**
    Randomly choose a client subset S
    Initialize $x_{i,0}^t = x_t$ for each client $i \in S$ concurrently
    for every client i in S **do**
        for $k = 0, 1, ..., K - 1$ **do**
            Calculate an objective approximation $g_{i,k}^t$ of $\nabla F_i(x_{i,k}^t)$
            Update $x_{i,k+1}^t$ = CLIENTOPT$(x_{i,k}^t, g_{i,k}^t, \eta_l, t)$
        end for
        Estimate the difference $\Delta_i^t = x_{i,K}^t - x_t$
    end for
    Compute the aggregated difference $\Delta_t = \frac{1}{|S|} \sum_{i \in S} \Delta_i^t$
    Update the global model x_{t+1} = SERVEROPT$(x_t, -\Delta_t, \eta, t)$
end for

Explanation of FedOPT Algorithm: Algorithm 1 demonstrates the FEDOPT algorithm, which has been designed to securely and efficiently carry out federated learning processes. It executes iteratively over a predetermined number of rounds, T, over which an arbitrary number of clients are selected, their specific models are configured, and these models get revised utilizing unbiased estimates in every round. The algorithm includes the initial global model x_0 and both client and server optimizers (CLIENTOPT and SERVEROPT) at the starting point of every process. Every round, a random subset S of clients is chosen. The method employs the present global model x_t to establish the local model xi for all clients i in subset S. Afterward, each client utilizes its own local data to perform local calculations over a predetermined number of iterations K. Using $x_{i,k}^t$, the client determines the desired estimation $g_{i,k}^t$, which represents the gradient of the loss function. The CLIENTOPT function is subsequently utilized iteratively to update the local model employing these mathematical estimates. Upon the execution of the local updates, all clients calculate the distinction, known as Δ_i^t, between the current global model x_t and the updated local model $x_{i,k}^t$. To determine the aggregated difference Δ_t, these variations are combined collectively for all clients in subset S. Eventually, the SERVEROPT function utilizes the aggregated difference Δ_t to modify the global model; this method repeats for T rounds. In brief, FedOPT provides decentralized optimization in federated learning contexts through the continuous improvement of local models for specific groups of clients, while modifications are combined to continually enhance the global model.

5 Results

5.1 Evaluation Metrics

The confusion matrix may be employed to construct an assortment of classification quality metrics that highlight various features of the classification. Nonetheless, a single value that summarizes the classification's accuracy is typically desired. Regretfully, evaluating categorization accuracy can prove to be a more demanding and intricate undertaking than it initially seems [46]. Therefore, to evaluate the performance of a classification model, one should use more than one metric to get a comprehensive idea of the efficacy of the model. All of them provide different perspectives on some facets of the performance of a model. Some commonly used metrics in machine learning are accuracy, precision, recall, and the F1-score. These metrics provide a way to measure model efficiency in correctly predicting the right class labels, handling class imbalances, and other important issues involving cost trade-offs between false positives and false negatives. Deep knowledge of these metrics is required in choosing models and fine-tuning them for their capacity to perform consistently well under real-world situations.

- **Accuracy:** Accuracy gives the rough estimate of the frequency of correctness of the classifier and is the measure of the ratio of correctly classified instances against total instances. Equation 6 illustrates how it is computed:

$$\text{Accuracy} = \frac{T.Positive + T.Negative}{T.Positive + T.Negative + F.Positive + F.Negative} \tag{6}$$

- **Precision:** Precision evaluates the quantity of true positive forecasts relative to all positive forecasts generated by the model, with high precision indicating a minimal false positive rate. Precision is calculated using Eq. 7:

$$\text{Precision} = \frac{T.Positive}{T.Positive + F.Positive} \tag{7}$$

- **Recall:** Recall quantifies the true positive forecasts in comparison to all real positive instances, playing a vital role in scenarios where missing a positive instance incurs substantial consequences. Recall is given by Eq. 8:

$$\text{Recall} = \frac{T.Positive}{T.Positive + F.Negative} \tag{8}$$

- **F1-Score:** The F1-score corresponds to the harmonic mean of precision and recall, presenting a unified measure that addresses both aspects, especially beneficial in cases of imbalanced class distribution by accounting for false positives and false negatives. F1-Score is depicted in Eq. 9:

$$\text{F1-score} = 2 \times \frac{Precision \cdot Recall}{Precision + Recall} \tag{9}$$

In the above equations,

- T. Positive: True Positive, which denotes that the system accurately anticipates the label to be positive.
- T. Negative: True Negative, which indicates that the system accurately anticipates the label to be negative.
- F. Positive: False Positive, which implies that the system interprets something as positive even though it is actually negative
- F. Negative: False Negative, which suggests that although it is a positive, the system classifies it as a negative.

5.2 *Hyperparameter Optimization*

A machine learning method's efficacy is greatly impacted by hyperparameters, which are external settings that have been established before the training phase and govern the learning process. Hyperparameters, as opposed to training parameters, must be specified in advance and can have a substantial effect on an algorithm's accuracy, efficiency, and generalization capabilities. Choosing the best hyperparameters demands experimentation as well as tuning to determine configurations that balance the complexity of the model and performance across multiple datasets. Appropriately tuned hyperparameters improve predicted accuracy, computational efficiency, and the ability of the model to generalize well to novel data, thus rendering them critical for designing efficient machine learning applications. In the remainder of this section, we have covered the hyperparameters that were employed throughout this, study:

- **Hyperparameters for LinearSVC**
  - Regularization Parameter (C): The penalty parameter for the error term in LinearSVC is denoted by the C parameter. It addresses the trade-off involving minimizing the complexity of the model and obtaining a low rate of error on the training set. While an increased value of C attempts to classify every training instance properly by providing the model more leeway to adapt the data, a

lesser value of C smoothes the decision surface. The aim of SVC is to solve the optimization problem shown in Eq. 10:

$$\min_{\omega,\beta,\zeta}\left(\frac{1}{2}\|\boldsymbol{\omega}\|^2 + C\sum_{i=1}^{n}\zeta_i\right) \tag{10}$$

where, $y_i(\boldsymbol{\omega}\cdot\mathbf{x}_i+\beta)\geq 1-\zeta_i,\quad \zeta_i\geq 0.$
Here:

* $\boldsymbol{\omega}$ refers to the specific weight vector that states the hyperplane.
* β denotes the bias term.
* ζ_i are the slack variables allowing the training samples to be misclassified.
* C indicates the regularization parameter.
* y_i are the class labels.
* $\mathbf{x}_i$ are the feature vectors.

A C value of 0.001 was discovered to yield the best performance on the dataset following tuning of the hyperparameter. The decision to opt for this number was based on its potential to accomplish the highest level of classification performance by striking an equilibrium between the degree of complexity and accuracy.

- **Hyperparameters for FedOpt Algorithm**
  - Coefficient for the First Moment: For the FedOpt algorithm's first moment estimates, the parameter in question represents the exponential decay rate. It regulates the extent to which the algorithm takes into account previous gradients during the update phase. A value near 1 indicates a higher weighting of previous gradients. The equation in this regard is presented in Eq. 11:

$$m_t = \beta_1\cdot m_{t-1} + (1-\beta_1)\cdot g_t \tag{11}$$

where:

* m_t: Updated 1st moment (mean) at time step t.
* m_{t-1}: Previous 1st moment.
* g_t: At time step t, the gradient of the loss function with respect to the model parameters.

We have balanced the use of historical gradients (m_{t-1}) and recent gradient information (g_t) in our FedOpt implementation by selecting a value of 0.9. This will be a deliberate weighting scheme to ensure the robust convergence and stability of the optimization process.

  - Coefficient for the second moment: This is the exponential decay rate for the second-moment estimations in the FedOpt algorithm. It helps to smooth down gradients when they are sparse or noisy. Setting it to a value closer to 1 gives more weight to prior squared gradients. The equation in this regard is presented in Eq. 12:

$$v_t = \beta_2 \cdot v_{t-1} + (1 - \beta_2) \cdot g_t^2 \tag{12}$$

where:

* v_t: Updated 2nd moment (variance) at time step t.
* v_{t-1}: Previous 2nd moment.
* g_t^2: Element-wise square of the gradient g_t.

Setting the value to 0.999 prioritizes the influence of recent squared gradients (g_t^2) in determining parameter updates (v_t). This high momentum factor effectively smooths out update steps, fostering consistent and reliable model optimization across successive training epochs.

- Epsilon: Epsilon is chosen as 1e-8 to avoid division by zero as well as to improve the numerical consistency of the FedOpt algorithm. By adding this small constant to the denominator when computing updates, epsilon ensures that the algorithm does not encounter numerical instabilities, especially when the second-moment estimate (v_t) approaches zero. This small value is standard practice in Adam-like optimizers to maintain robustness during training.
- Learning Rate: The global model's modifications to parameters are performed utilizing the Learning Rate as the dimension of the step. Depending upon the gradients, it establishes how rapidly or gradually the model shifts its parameters over time. When the learning rate is higher, there is an increased likelihood of overshooting the optimum solution, even with more rapid updates. Conversely, lower learning rates entail slower yet possibly more reliable convergence. In the context of FedOpt, the update rule for model parameters ϑ is typically defined as Eq. 13:

$$\theta_{t+1} = \theta_t - \eta \cdot \frac{m_t}{\sqrt{v_t} + \epsilon} \tag{13}$$

where:

* θ_t and θ_{t+1}: Parameters of the model at time steps t and $t + 1$, in that order.
* η: Learning rate, determining the step size of the update.
* m_t: Estimate of the first moment (mean) of gradients with bias correction applied.
* v_t: Estimate of the second moment (variance) of gradients with bias correction applied.
* ϵ: Small constant that prevents division by zero.

The learning rate, 0.001 in this case, is adopted to control the magnitude of parameter adjustments so that a proper balance between the speed of convergence and stability is achieved. This calibrated learning rate allows for controlled

Table 3 Optimal values of hyperparameters

Federated algorithm	Model	Regularization parameter (C)	Coefficient for the first moment	Coefficient for the second moment	Epsilon	Learning rate
FedOpt	SVM	0.001	0.9	0.999	1e-8	0.001

updates, (θ_t), under gradient estimations, ($\hat{m}_t$ and $\hat{v}_t$), ensuring model performance optimization without the risk of overshooting optimal parameter values. The Table 3 displays the hyperparameters at their optimal values.

5.3 *Performance Evaluation*

The FedOpt-SVM obtained remarkable metrics on the training set, with 95.62% accuracy, 94.44% precision, 100% recall, and 97.14% F1-score. These results underscore the model's precision in minimizing false positives and effectively identifying all positive cases. Evaluation on the test set yielded a commendable 87.50% accuracy and 87.50% precision, with a perfect 100% recall and 93.33% F1-score. These results therefore confirm the generalization ability of the SVM model in ensuring a reliable diagnosis of rheumatism in unseen data, made possible through the innovative FedOpt framework for collaborative model training. This study demonstrates the SVM model's potential to advance clinical decision-making in rheumatology and enhance personalized healthcare outcomes through cutting-edge machine learning methodologies.

The performance of the federated model implemented in this study is presented in Table 4.

Table 4 Performance metrics of federated learning algorithms

Federated algorithm	Model	Test data				Train data			
		Accuracy	Precision	Recall	F1-Score	Accuracy	Precision	Recall	F1-Score
FedOpt	SVM	87.50%	87.50%	100%	93.33%	95.62%	94.44%	100%	97.14%

5.4 *Analysis of the Confusion Matrices*

A confusion matrix is a performance evaluation tool applied to address the problems associated with classification in machine learning. It represents a table arrangement that allows comparison of the real and anticipated numbers concerning some cases to be visualized by the degree to which an algorithm is performing. The predictions are then grouped based on whether they are true positives, true negatives, false positives, or false negatives using the matrix. This split makes it easier to see where the model is failing and how accurately it forecasts each dataset class. Figure 2 visualizes the training and testing set confusion matrices for this study.

5.4.1 Training Set Confusion Matrix

The training set confusion matrix showcases a resounding success in the classification task. The model has phenomenally distinguished between the two classes, achieving an almost perfect separation. Out of the 41 samples in class "0," a remarkable 34 were correctly classified, illustrating the model's impressive precision. While the model did misclassify 7 samples, an extraordinary performance overshadows this minor discrepancy.

The model's ability to correctly classify all 119 samples from class "1" is an outstanding achievement. This flawless accuracy for class "1" highlights the model's impeccable capability in recognizing and correctly identifying this class. In effect, this means that the model can provide exemplary performance on the training set, thus raising a high standard for model performance.

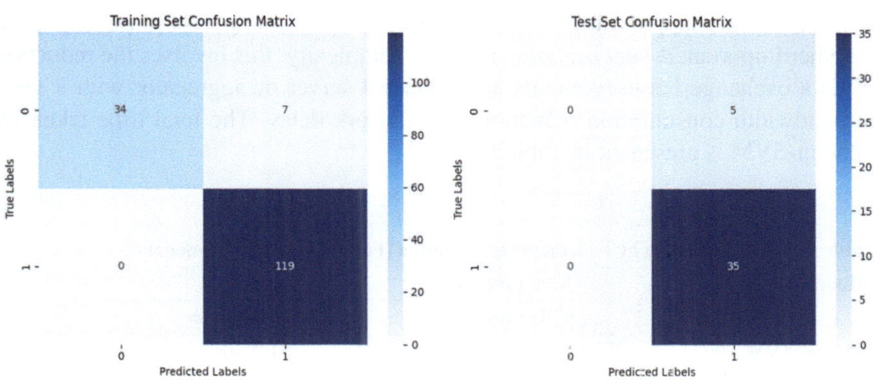

Fig. 2 Confusion matrices of FedOpt-SVM

5.4.2 Test Set Confusion Matrix

Turning to the test set confusion matrix, the model consistently demonstrates a good result. While there were no true negatives and 5 false negatives for class "0," it is important to note that generalization to unseen data is difficult. It is amazing that the model correctly classified 35 out of 35 samples in class "1" on the test set which demonstrates its robustness and reliability.

This outstanding classification of class "1" demonstrates that this model has a great generalization ability as well as an incredible ability to remain accurate even on unseen data. The test set performance which points out areas for further improvement also underscores once again just how immensely strong and competent this mode is in dealing with complex classification tasks.

5.5 Complexity Analysis

In computer science, complexity evaluation is a procedure to measure efficiency in algorithms on increasing input sizes. It basically deals with two major factors: space complexity, that is, the amount of memory used by an algorithm under similar conditions, and time complexity, meaning the amount of time taken by an algorithm to run as the input increases. Complexity analysis in Federated Learning can be used to estimate the computational and communication burdens incurred by modeling machine learning algorithms across distributed servers or devices. On its part, the complexity analysis for FedOpt looks directly at many factors that are paramount for its operation. In other words, computational loads at the stages of local aggregation and model update for any additional computational overhead introduced by the optimization techniques and the power needed for processing have to be estimated. Another important aspect is communication complexity: this involves the reduction of data exchange between clients and a central server or aggregator with a view to bandwidth consumption reduction and network delay. The total time taken by FedOpt-SVM is presented in Table 5.

Table 5 Total time taken by Federated Optimization (FedOpt) with SVM model

Model	Total time taken	
	CPU time	Wall time
FedOpt-SVM	59.6 ms	179 ms

6 Conclusion

This paper proposes a unique solution to employ federated learning as a technique to improve the recognition of rheumatic diseases while at the same time not compromising the privacy rights of the users. Federated learning allows overall data trends to be obtained without the necessity for data to be centralized, hence maintaining the efficiency and effectiveness of the detection system. The present study reveals how federated learning is capable of harnessing a large amount of strength to detect rheumatism within the criterion, thus providing a robust system of privacy-preserving, precise, and effective diagnostic techniques. Thus, the problem of data privacy was solved with the help of federated learning, which proposes a way to collect data from many sources without their consolidation. This approach, besides safeguarding user privacy, also greatly improved the stability and versatility of the constructed detection model. Thus, normalization of data and selection of the most significant features in the dataset allowed us to achieve the best conditions for the use of machine learning. These techniques were crucial in defining the nature of the data, and in choosing the features that would be appropriate for the model, hence making it much more efficient and reliable. Using FedOpt as the framework for optimization, the SVM-based model displayed remarkable performance. On the training set, it had an accuracy of 95.62%, precision of 94.44%, recall of 100%, and an F1-score of 97.14%. In addition, on the test set, it recorded impressive accuracy, recall, precision, and F1-scores of 87.50%, 87.50%, 100%, and 93.33%, respectively. These findings attest to the efficiency of the model in minimizing false positives while successfully detecting each of the true positive cases; thus, the ability to offer accurate diagnosis in unseen samples. As demonstrated by the success of the proposed approach, federated learning presents a powerful avenue for dramatically changing clinical decision-making for rheumatological disorders in the future and a new standard for privacy-preserving rheumatism diagnosis systems. Furthermore, the application of federated learning made it possible to prevent the centralization of user data, which allowed for privacy to be protected throughout the model's creation and implementation processes. Such an approach is essential in the modern world, where data is a valuable commodity and privacy is crucial, especially when it comes to personal data. Thus, by ensuring user privacy while achieving a high accuracy rate in disease detection, we present a solid foundation that can be developed and generalized for other medical conditions as well, which will lead to an increase in the security and effectiveness of healthcare systems around the globe.

References

1. K.P. Gasnick, Rheumatism: Meaning, disease groups, symptoms, effects (2023). Accessed 13 April 2024. https://www.verywellhealth.com/rheumatism-8386785
2. C.A. Mancuso, R. Duculan, D. Jannat-Khah, M. Barbhaiya, A.R. Bass, B. Mehta, Rheumatic disease-related symptoms during the height of the covid-19 pandemic. HSS J.® **16**(1_suppl), 36–44 (2020)
3. R.J. Black, M. Cross, L.M. Haile, G.T. Culbreth, J.D. Steinmetz, H. Hagins, J.A. Kopec, P.M. Brooks, A.D. Woolf, K.L. Ong et al., Global, regional, and national burden of rheumatoid arthritis, 1990–2020, and projections to 2050: a systematic analysis of the global burden of disease study 2021. Lancet Rheumatol. **5**(10), e594–e610 (2023)
4. J. Tian, D. Zhang, X. Yao, Y. Huang, Q. Lu, Global epidemiology of systemic lupus erythematosus: a comprehensive systematic analysis and modelling study. Ann. Rheumat. Dis. **82**(3), 351–356 (2023)
5. B. Grygiel-Gorniak, E. Masiero, B.C. Nevaneeth, M.M. Jojy, Rheumatic diseases in reproductive age—the possibilities and the risks. Reprod. Sci. **30**(1), 111–123 (2023)
6. S.M. Bramanti, V. Manippa, A. Babore, A. Dilillo, A. Marcellino, V. Martucci, S. Mallardo, S. Isoldi, S. Bloise, M. Sanseviero et al., Comparing parental distress and children's difficulties between parents of children with rheumatic diseases and parents of healthy children in families facing the covid-19 pandemic. Current Psychol. **42**(23), 20114–20123 (2023)
7. T. Xie, M. Masroor, X. Chen, F. Liu, J. Zhang, D. Yang, C. Liu, M. Xiang, Rheumatism as a cause of cardiac hemangioma: a rare case report and review of literature with special focus on etiology. BMC Cardiovasc. Disord. **23**(1), 203 (2023)
8. D. Mistry, M.R. Tonmoy, M.S. Anower, A.S.M.T. Hasan, Federated transfer learning for vision-based fall detection, in *Proceedings of the 2nd International Conference on Big Data, IoT and Machine Learning* ed. by M.S. Arefin, M.S. Kaiser, T. Bhuiyan, N. Dey, M. Mahmud (Springer Nature Singapore, Singapore, 2024), pp. 961–975
9. J. Bianchi, A.C. de Oliveira Ruellas, J.R. Goncalves, B. Paniagua, J.C. Prieto, M. Styner, T. Li, H. Zhu, J. Sugai, W. Giannobile, et al., Osteoarthritis of the temporomandibular joint can be diagnosed earlier using biomarkers and machine learning. Scientific Rep. **10**(1), 8012 (2020)
10. M. Luque-Tévar, C. Perez-Sanchez, A.M. Patiño-Trives, N. Barbarroja, I. Arias de la Rosa, M.C. Abalos-Aguilera, J.A. Marin-Sanz, D. Ruiz-Vilchez, R. Ortega-Castro, P. Font et al., Integrative clinical, molecular, and computational analysis identify novel biomarkers and differential profiles of anti-tnf response in rheumatoid arthritis. Front. Immunol. **12**, 631662 (2021)
11. J. Wen, L. Wan, X. Dong, Novel peripheral blood diagnostic biomarkers screened by machine learning algorithms in ankylosing spondylitis. Front. Genet. **13**, 1032010 (2022)
12. Y. Liu, H. Jiang, T. Kang, X. Shi, X. Liu, C. Li, X. Hou, M. Li, Platelets-related signature based diagnostic model in rheumatoid arthritis using wgcna and machine learning. Front. Immunol. **14**, 1204652 (2023)
13. F. Liu, J. Ye, S. Wang, Y. Li, Y. Yang, J. Xiao, A. Jiang, X. Lu, Y. Zhu, et al., Identification and verification of novel biomarkers involving rheumatoid arthritis with multimachine learning algorithms: an in silicon and in vivo study, Mediators of Inflammation (2024)
14. A. Vanier, J.S. Smolen, C.F. Allaart, R. Van Vollenhoven, P. Verschueren, N. Vastesaeger, S. Saevarsdottir, K. Visser, D. Aletaha, B. Combe et al., An updated matrix to predict rapid radiographic progression of early rheumatoid arthritis patients: pooled analyses from several databases. Rheumatology **59**(8), 1842–1852 (2020)
15. M. Kalweit, U.A. Walker, A. Finckh, R. Müller, G. Kalweit, A. Scherer, J. Boedecker, T. Hügle, Personalized prediction of disease activity in patients with rheumatoid arthritis using an adaptive deep neural network. PLoS One **16**(6), e0252289 (2021)
16. H. Matsuo, M. Kamada, A. Imamura, M. Shimizu, M. Inagaki, Y. Tsuji, M. Hashimoto, M. Tanaka, H. Ito, Y. Fujii, Machine learning-based prediction of relapse in rheumatoid arthritis patients using data on ultrasound examination and blood test. Sci. Rep. **12**(1), 7224 (2022)

17. R. Queiro, D. Seoane-Mato, A. Laiz, E. Galindez Agirregoikoa, C. Montilla, H.S. Park, J.A.P. Tasende, J.J.B. Baute, B. Joven Ibáñez, E. Toniolo, et al., Moderate-high disease activity in patients with recent-onset psoriatic arthritis—multivariable prediction model based on machine learning. J. Clin. Med. **12**(3), 931 (2023)
18. J. Batool, A. Hakim, J. Jabeen, A diagnostic tool for rheumatoid arthritis and disease activity levels using machine learning classifiers. J. Comput. & Biomed. Inf. (2024)
19. K. Üreten, H. Erbay, H.H. Maraş, Detection of hand osteoarthritis from hand radiographs using convolutionalneural networks with transfer learning. Turk. J. Electr. Eng. Comput. Sci. **28**(5), 2968–2978 (2020)
20. G.S. Mate, A.K. Kureshi, B.K. Singh, et al., An efficient cnn for hand x-ray classification of rheumatoid arthritis. J. Healthcare Eng. **2021** (2021)
21. K. Üreten, H.H. Maraş, Automated classification of rheumatoid arthritis, osteoarthritis, and normal hand radiographs with deep learning methods. J. Dig. Imag. **35**(2), 193–199 (2022)
22. S. Srinivasan, S. Gunasekaran, S.K. Mathivanan, P. Jayagopal, M A. Khan, A. Alasiry, M. Marzougui, A. Masood, A framework of faster crnn and vgg16-enhanced region proposal network for detection and grade classification of knee ra. Diagnostics **13**(8), 1385 (2023)
23. W. Shuai, X. Wu, C. Chen, E. Zuo, X. Chen, Z. Li, X. Lv, L. Wu, C. Chen, Rapid diagnosis of rheumatoid arthritis and ankylosing spondylitis based on fourier transform infrared spectroscopy and deep learning. Photodiagnosis Photodyn. Therapy **45**, 103885 (2024)
24. C. Ye, S. Cai, G. Shen, H. Guan, L. Zhou, Y. Hu, W. Tu, Y. Chen, Y. Yu, X. Wu et al., Clinical features of rheumatic patients infected with covid-19 in Wuhan, China. Ann. Rheum. Dis. **79**(8), 1007–1013 (2020)
25. A. Tripathy, N. Swain, B. Gupta, The covid-19 pandemic: an increased risk of rheumatoid arthritis. Future Virol. **16**(6), 431–442 (2021)
26. S.M. Burns, Z. Icten, T. Honda, J. Manjourides et al., A machine learning approach to identify predictors of severe covid-19 outcome in patients with rheumatoid arthritis. Pain Phys. **25**(8), 593 (2022)
27. N. Ferdous, M. Islam, M. Rahman, A. Zaman, F. Nazrul, Clinicodemographic characteristics of covid-19 infection among patients with rheumatic diseases. J. Bangladesh Coll. Phys. & Surg. **41**(3) (2023)
28. Q. Wang, Y. Wu, L. Wu, C. Lu, Y. Ke, Y. Wang, L. Gu, Y. Shen, W. Tan, Dynamics of sars-cov-2 igg antibodies and neutralising antibodies in rheumatic and musculoskeletal diseases patients with covid-19. Clinical and Experimental Rheumatology (2024)
29. D. Kim, G. Lee, S. Oh, Toward privacy-preserving text embedding similarity with homomorphic encryption, in *Proceedings of the Fourth Workshop on Financial Technology and Natural Language Processing (FinNLP)* (2022), pp. 25–36
30. W. Lu, A.L. Varna, M. Wu, Confidentiality-preserving image search: a comparative study between homomorphic encryption and distance-preserving randomization. IEEE Access **2**, 125–141 (2014)
31. A.P. Brown, C. Borgs, S.M. Randall, R. Schnell, Evaluating privacy-preserving record linkage using cryptographic long-term keys and multibit trees on large medical datasets. BMC Med. Inf. Dec. Mak. **17**, 1–7 (2017)
32. J. Flemings, M. Annavaram, Differentially private knowledge distillation via synthetic text generation (2024). arXiv:2403.00932
33. J. Andrew, R.J. Eunice, J. Karthikeyan, An anonymization-based privacy-preserving data collection protocol for digital health data. Front. Public Health **11**, 1125011 (2023)
34. D. Mistry, M.F. Mridha, M. Safran, S. Alfarhood, A.K. Saha, D. Che, Privacy-preserving onscreen activity tracking and classification in e-learning using federated learning. IEEE Access **11**, 79315–79329 (2023). https://doi.org/10.1109/ACCESS.2023.3299331
35. S. Boo, H. Oh, E.S. Froelicher, C.-H. Suh, The data file of 200 patients with rheumatoid arthritis. Dataset (2017). https://doi.org/10.1371/journal.pone.0176291.s001
36. G. Dudek, S. Sakowski, O. Brzezińska, J. Sarnik, T. Budlewski, G. Dragan, M. Poplawska, T. Poplawski, M. Bijak, J. Makowska, Machine learning-based prediction of rheumatoid arthritis with development of acpa autoantibodies in the presence of non-hla genes polymorphisms. Plos one **19**(3), e0300717 (2024)

37. B.S. Diba, J.D. Plabon, M.M. Rahman, D. Mistry, A.K. Saha, M. Mridha, Explainable federated learning for privacy-preserving bangla sign language detection. Eng. Appl. Artif. Intell. **134**, 108657 (2024)
38. M.G. Kibriya, D. Mistry, D. Mistry, M.T. Rahman, S.H. Ovi, A. Tabassum, S. Hossain, Privacy-preserving vision-based detection of pox diseases using federated learning, in *Data-Driven Clinical Decision-Making Using Deep Learning in Imaging* (Springer Nature Singapore, Singapore, 2024), pp. 105–122
39. A. Tabassum, S.H. Ovi, S. Hossain, M.R. Tonmoy, M.S.H. Shovon, M.R. Hussein, D. Mistry, Privacy preserving breast cancer prediction with mammography images using federated learning, in *Data-Driven Clinical Decision-Making Using Deep Learning in Imaging* (Springer Nature Singapore, Singapore, 2024), pp. 227–245
40. D. Mistry, J.D. Plabon, B.S. Diba, S. Mukta, M.F. Mridha, Federated learning-based architecture for personalized next emoji prediction for social media comments. IEEE Access 1–1 (2024). https://doi.org/10.1109/ACCESS.2024.3448470
41. M.S.H. Shovon, Z. Sultana, J. Shin, M.A. Hamid, D. Mistry, Incorporating residual connections into a multi-channel cnn for lung cancer detection in digital pathology, in *Data-Driven Clinical Decision-Making Using Deep Learning in Imaging*. (Singapore, Springer Nature Singapore, 2024), p.189
42. E. Jafarigol, T.B. Trafalis, A distributed approach to meteorological predictions: addressing data imbalance in precipitation prediction models through federated learning and gans. Comput. Manag. Sci. **21**(1), 22 (2024)
43. P. Kairouz, H.B. McMahan, B. Avent, A. Bellet, M. Bennis, A.N. Bhagoji, K. Bonawitz, Z. Charles, G. Cormode, R. Cummings, et al., Advances and open problems in federated learning. Found. Trends® Mach. Learn. **14**(1–2), 1–210 (2021)
44. M. Asad, A. Moustafa, T. Ito, Fedopt: towards communication efficiency and privacy preservation in federated learning. Appl. Sci. **10**, 2864 (2020)
45. S. Reddi, Z. Charles, M. Zaheer, Z. Garrett, K. Rush, J. Konečný, S. Kumar, H.B. McMahan, Adaptive federated optimization (2020)
46. G.M. Foody, Challenges in the real world use of classification accuracy metrics: from recall and precision to the matthews correlation coefficient. Plos one **18**(10), e0291908 (2023)

Enhancing Gall Bladder Cancer Detection by Privacy-Preserving Federated Learning: A Comprehensive CSV-Based Classification Approach

Bidita Sarkar Diba, Nishat Tasnim, Mehjabin Hossain, Aloke Kumar Saha, Md. Golam Kibriya, and Moshiur Rahman Tonmoy

Abstract Gallbladder cancer (GBC) is highly aggressive and clinically unpredictable, making the prognosis and treatments quite complex. This study aims to investigate whether an existing federated learning (FL) framework, together with the FedProx algorithm, can improve predictive analysis for GBC patients. We used CT scan data and integrated radiomic features with clinicopathological data to create an accurate prognostic model. The proposed FL framework enabled by the FedProx algorithm consists of a simply integrated radiomic-clinical dataset providing information accuracy, precision, recall, and F1-score based on the training and testing dataset of the federated K-nearest neighbors' (FedKNN) model. Our results show that the FedProx algorithm, through the incorporation of FedKNN, can perform and attain an accuracy of approximately 92.86% on the training set and 89.66% on the test set, demonstrating the model's ability to predict the outcomes of gallbladder cancers and its applicability in combining various data sources for enhanced predictive capabilities in oncology.

B. S. Diba · N. Tasnim · A. K. Saha
Department of CSE, University of Asia Pacific, Dhaka, Bangladesh
e-mail: bidita@uap-bd.edu

N. Tasnim
e-mail: 21201149@uap-bd.edu

A. K. Saha
e-mail: aloke@uap-bd.edu

M. Hossain
Advanced Machine Intelligence Research Lab, Dhaka, Bangladesh
e-mail: 21201004@uap-bd.edu

Md. G. Kibriya
Department of EEE, Pabna University of Science and Technology, Pabna, Bangladesh

M. R. Tonmoy (✉)
Department of Computer Science & Engineering, Bangladesh University of Business and Technology, Pabna, Bangladesh
e-mail: moshiurtonmoy@bubt.edu.bd

M. F. Mridha and N. Dey (eds.), *Federated Learning in Health Care Technology*, Studies in Computational Intelligence 1216, https://doi.org/10.1007/978-981-96-8353-6_9

Keywords Gallbladder cancer · Federated learning · Federated proximal · Federated K-nearest neighbors · FedKNN

1 Introduction

Gallbladder cancer (GBC), though a rare cancer, is one of the most malignant cancers among the biliary tract cancers [41]. It tends to be diagnosed at an advanced stage, causing the deterioration of an already bleak prognosis with a low survival rate. GBC is also a great threat since it has no signs or symptoms that can lead to early identification; rather, it is normally detected late, at the same time as gallstones or cholecystectomy. Although considered a minor cancer, GBC exerts a significant influence and is associated with a significant amount of morbidity and mortality globally. It accounts for 7% of all cancer deaths and about 3.5 million disability-adjusted life years [21, 22]. The development of GBC is further worsened by poor screening procedures, which has led to early detection of the disease. Sadly, according to Rawla et al. [42], the majority of GBC patients are diagnosed at the advanced or metastatic stage, which negatively impacts the prognosis. Five-year relative survival of only 19% further emphasizes the need for early detection and better outcomes [16]. In order to decrease the burden of GBC on health systems and societies, it is essential to identify the etiological factors and risk factors of GBC.

The clinical relevance of GBC in the USA alone is noteworthy; 12,350 new cases and 4,530 deaths in [2]. Though a rare cancer compared to others, GBC manifests the common symptoms of jaundice, pain, and fever at the final stages of the disease. One of the issues that are encountered in diagnosing GBC histopathologically is that the disease, in some instances, may be superficial involving only the mucosa [50]. While a significant prognostic impact is noted when these lesions are identified at an early stage, however, most patients present with symptoms at an advanced stage of the disease, which restricts the potential for curative surgery. Various prognostic factors, such as the presence of gallstones, further complicate the clinical landscape of GBC. The development of GBC in less than 1% of patients with cholelithiasis, although often associated, highlights the complex pathophysiological mechanisms involved [46]. Treatment options range from curative surgical interventions for suspected mucosal lesions to palliative procedures for advanced, unresectable, metastatic, or recurrent cases. However, the effectiveness of available treatment options remains limited, necessitating innovative approaches to the management of GBC [29].

In this process, the first motivation comes from noticing that patients are diagnosed with GBC at a later stage; the therapeutic management of the disease is restricted; and the five-year survival rate is low. The current diagnostic model is a scenario of late diagnosis and meager screening programs; thus, there is a need to advance a new model to boost the early diagnosis rate and hence the results of its treatment [42]. The identification of GBC in its early stages is of great importance due to the lack of therapeutic possibilities and the low survival rate during the later stages of the disease.

Cooperation between data protection and medical research, particularly GBC detection, was addressed significantly by Privacy-Preserving Federated Learning (FL). FL enables co-training from distributed databases without assembling patient data, breaching patients' privacy and rights while fostering the development of generalizable AI learning models [49]. FL makes it possible to gather datasets from various institutions or geographical areas, making the process of the GBC detection models more flexible and scalable compared to data repositories.

Besides, the integration of FL with the GBC detection domain obeys the tendencies in the field of health, relying on individual and dedicated data-driven approaches. FL also supports the development of reference and situational models as necessary for managing the GBC spectrum in various patient populations. Such an approach not only enhances the efficacy of detection of GBC but also the likelihood of early diagnoses and treatments leading to proper management strategies and increased chances of better results among the patients. The call for FL on GBC detection research stems from the ethical and legal issues always associated with data sharing in the medical field [12]. This decentralized nature means that the patient's records do not escape the confines of the institution, thus safeguarding personal details from leakage and infringement of the law. First, FL guarantees patients' privacy and confidentiality, which establishes trust and cooperation among the stakeholders concerned with the consistent sharing of knowledge and experience in striving for the identification of better GBC diagnostic methods.

With FL, researchers can fine-tune accurate and portable ML models and secure patients' data. These integrated and collaborative models can change the way of diagnosis and management of GBC, improving patients's outcomes and reducing the costs of healthcare services. This research work is therefore multi-fold and seeks to enhance the classification of GBC using a classification system of CSV and the privacy-preserving FL technique. If FL is implemented into the GBC detection system, several improvements are anticipated, ranging from improvements in the performance of the model to improvements in data privacy and combined datasets. The first objective is to define a feasible and accurate paradigm of GBC. The conventional screening test involves the use of medical data that are read by radiologists, and this may be sensitive and variable. Therefore, FL will assist in gathering the intelligence of the dispersed healthcare centers and utilize this data set to build a single GBC detection model that will also be scalable. On the other hand, FL assists in the decentralization of patients' data and, hence, privacy in the training of multi-agency collaborative models [35]. The method employed in the study also strengthens the GBC detection model concurrently as it optimizes patient samples to make their model more relevant and widely applicable.

The integration of FL into the GBC detection framework results in improved conditions of privacy. This is because the model updates can be computed locally, with only model updates being exchanged. With FL, we would like to expand numerous inclusive datasets to mitigate the issue of data scarcity and imbalance in GBC identity research. Weak model capacity, a shortage of data samples, and intrinsic problems of minority classes are inherent in traditional machine learning and result in the creation of models with poor generalization capability. FL mitigates the above by training the models across multiple datasets that bring together information useful in generalizing on unseen data [24]. The main goal of this work is to incorporate the Federated Learning that Preserves Privacy features into a multilevel classification model for gallbladder cancer detection.

The main contribution of this paper includes:

- Employed the FedKNN model with the FedProx algorithm to predict gallbladder cancer survival outcomes with a high accuracy of 92.86% on the training set and 89.66% on the test set.
- Optimized hyperparameters for the FedKNN model, including those related to training and evaluation, which enhanced its performance in predicting gallbladder cancer outcomes.
- Utilized confusion matrix analysis to assess the model's performance, highlighting its high recall rate for positive cases and addressing precision and false positive issues.

The rest of the paper is structured as follows: Related Work explores methods and innovations for gallbladder cancer diagnostics utilizing deep learning and machine learning. Conventional Privacy-Preserving Methods address how to secure sensitive CSV dataset information. Methodology provides dataset description, preprocessing, the proposed framework, and the FedProx algorithm. Result and Analysis signifies experimental results, evaluation metrics, and comparisons to past research. Conclusion and Future Work highlights results, contributions, and directions for further research.

2 Related Work

Gallbladder cancers (GBCs) have a poor prognosis and are difficult to diagnose and cure. However, as a result of advancements in deep learning (DL) and machine learning (ML), there are genuinely remarkable possibilities for improving GBC diagnosis and prognosis. This section covers a wide range of techniques, including Bayesian networks, machine learning, deep learning, and their integration.

2.1 Deep Learning (DL) Models

DL models have been extensively studied to address a variety of issues pertaining to gallbladder disease and gallbladder cancer. A study used deep feature extraction based on a 3D-DenseNet model to construct a survival prediction model for GBC patients. Comparing the results of the laboratory tests, systemic treatments, and enhanced CT scans with multimodal medical data, a notable improvement in survival prediction accuracy was obtained over the previous models. The rarity of gallbladder cancer, however, as well as missing values in clinical and systemic therapy data, the need to process incomplete data through interpolation, the lack of disease information along the temporal dimension, and the change in the patient's disease course, all were identified as limitations [15]. Additionally, a Convolutional Neural Network (CNN) model for the automated diagnosis of gallbladder cancers was developed and evaluated in another study using information from stomach ultrasonography scans. The CNN version employed a Deep Learning (DL) classifier that was developed using a multiscale, second-order pooling-based methodology. This CNN model showed impressive sensitivity and specificity, highlighting its potential to support diagnosis, especially in resource-constrained environments. It is essential to understand certain limitations of the research, such as the dependence on data from a single center, limited patient numbers in particular subgroups (like polyps), absence of an accurate evaluation regarding CNN's influence on the fast diagnosis and prognosis of cancer of the gallbladder, and the unavailability of the study's data for public access [7]. Another study used a Generative Adversarial Network (GAN) framework to investigate CT results related to gallbladder cancer, showcasing several neural network designs specifically designed to improve picture quality and validate photos with reduced noise. This demonstrates how GANs can provide a dependable supplementary prognosis for the majority of gallbladder malignancies. The study used low-dose computed tomography (LDCT) of the majority of gallbladder tumors with different noise levels, distributions of artifacts, and dosages. Although the observation achieved excellent accuracy, it was unable to provide statistics on the size of the pattern or the particular dataset used Chang et al. [13]. Moreover, a deep neural community architecture that included local as well as global attention mechanisms was introduced to research that used the Gallbladder Cancer Ultrasound (GBCU) dataset to diagnose gallbladder cancer from ultrasound pictures in a novel way. The Radiology Transformer (RadFormer) model improved understanding in this area by not only enabling accurate identification of gallbladder cancer but also providing comprehensible justifications for decision-making. However, the limitations in identifying positive vital signs of malignancy in gallbladder cancer from ultrasound result in false negatives and incorrect forecasts [8]. In a different study, ultrasound of the gallbladder was examined using a Transfer Learning Deep Neural Network (DNN) Algorithm that integrated many CNN models, including VGG16, InceptionV3, ResNet152, and MobileNet. Particularly, MobileNet was utilized to simultaneously identify nine distinct types of gallbladder diseases, demonstrating the efficacy of deep learning in accurately identifying several problems, with MobileNet achieving the greatest accuracy. However, the examination also identified issues with inadequate quality and aberrations in ultrasonic scanning, which may further complicate the process of effectively separating intra-belly organs [34]. A brief review included significant improvements in gallbladder disease diagnosis using endoscopic ultrasound (EUS). The objective was to increase diagnostic specificity through the integration of artificial intelligence with deep learning (DL) and ResNet50 and InceptionV3 models. This marked a shift from machine learning to Deep Learning procedures. These procedures have significantly improved the accuracy of diagnosing specific gallbladder anomalies, such as GB wall thickening and protuberant lesions, using ultrasound pictures by utilizing contrast-enhanced endoscopic ultrasound (CH-EUS) and EUS [1].

Similar to this, another study used CT data from the TRECVID and UCF101_SBD datasets and 3D-DenseNet technology for a more effective diagnosis of GBC. Nevertheless, the potential of GBC diagnosis had been revealed in the context of the study, despite weak links such as limited data accessibility and poor model calibration. Of course, there are disadvantages, and one of them is the lack of information about gallbladder cancer, which is not suitable for establishing a medical imaging database. Also, study results were not compared with other testing methods such as ROC curve analysis, specificity, sensitivity, and accuracy [10]. In another related study, CNNs that were trained on CT scan data were used to differentiate between gallbladder cancer and benign pathology. The nearby liver parenchyma was incorporated during training to enhance the CNN's performance; thus, it may help in describing the gallbladder lesions. Nevertheless, the CNN was trained and tested on a small amount, as gallbladder cancer is not common in daily clinical practice, and the trial was conducted at a single center. The size of the dataset could be another disadvantage of CNN and might affect the level of generalization. In an attempt to fill this research gap, as stated in Derya et al. [18], the GBCNet, a CNN-based model designed for the identification of gallbladder cancer from USG pictures, was proposed. The problems of low quality and issues with shadows were solved, which showed higher accuracy in the results of GBCNet in comparison with radiologists and other models. Though these constraints are not fully addressed, the research does note the limits of CNN models in evaluating USG pictures, including problems like poor imaging quality, noise, and misalignment of views [9]. Last but not least, to enhance the result of surgery, a study on the XGC and GBC was carried out with the help of CT data and DL. The highest discriminating performance was achieved for the test dataset using a CT patch-based model with a residual convolutional neural network and using 5-fold cross-validation [23].

The study presents a new approach to selecting the correct interventions and avoiding unnecessary invasive procedures. However, the model is limited to a small patient population; hence, its applicability is still restricted.

2.2 *Machine Learning (ML) Models*

In a multicenter retrospective cohort study, a novel approach showcasing differentiation between gallbladder cancer and xanthogranulomatous cholecystitis before surgery was demonstrated using the random forest technique. The following nomogram offers physicians a useful diagnostic aid since it was effective in the differentiation process of the two conditions [6]. Furthermore, LASSO was employed during the ML algorithm to build the regulatory network of GBC using the distinct microRNAs. The AUC value and ROC curve for this method explained the high rate of accurate efficiency in the identification of unique or peculiar genes for gallbladder cancer [43]. In addition, multiple institutional studies for preoperative characteristics

(CART), also known as machine learning analysis, were utilized to divide gallbladder cancer patients into different groups. Using variables that were significantly associated with OS, the NRP was able to come up with risk groups that patients belong to Cotter et al. [17]. Also, in an attempt at survival prediction for prognosis focusing on a specific biomarker, a classifier was developed using radiomics with LASSO and Random Forest on CT scans. The proposed technique also offered viable differential diagnostic solutions by accurately estimating the survival rate of patients with gallbladder cancer based on the integration of clinical features [33].

Common limitations were seen when using machine learning algorithms to diagnose gallbladder disease. The quick development of deep learning methods is one of them, which can underrepresent more recent studies. Further limiting direct outcome comparisons are heterogeneous methodology and performance metrics. Due to deep learning models' strong reliance on the quantity and quality of data, generalizability is limited when small, homogeneous datasets are used. Moreover, insufficient details about datasets and model parameters impede reproducibility and transparency.

2.3 *Optimization Algorithms*

In more investigation into the diagnosis and prognosis of Gallbladder Carcinoma, an optimization algorithm utilizing BPNN and GA for combined detection of tumor markers in gallbladder cancer was conceptualized. The BPNN and GA approach achieved a remarkable accuracy of 94.18% for detecting CA242 and CA199 [14]. Furthermore, for the classification of gallbladder lesions, a hybrid solution was employed in optimal radiological gallbladder lesion characterization by combining visual assessment with CT-based radiomics to attain optimal characterization of the gallbladder lesions through the use of ridge regression with radiomic characteristics extracted from the CT data. This work established how the technique of ridge regression can enhance healthcare decision-making and diagnostic accuracy where gallbladder lesions are concerned [51].

These optimization algorithms for diagnosing gallbladder cancer are restricted by their small stony-size and single-center study set, rate uncertainty about the effect of bile, sludge, and stones on radiomics analysis, the applicability of the algorithm for the diagnosis of gallbladder malignancy only, and the unsuitability for other malignant neoplasms.

2.4 *Combination of ML and DL*

Basically, for diagnosing diseases, the Support Vector Machine (SVM) proved its efficiency compared to the other research ML and DL techniques by acquiring the maximum accuracy rate for gallbladder disorders [3]. Logistic regression was used to evaluate the metabolomic expression data involving gallstone disease prediction in a particular study. The study demonstrated the possibility of logistic regression in disease prediction by producing encouraging results in gallstone disease prediction [4]. In a radiomics-based investigation, the ability of radiomics models to predict serosal involvement in gallbladder cancer was examined by combining deep learning and machine learning algorithms based on contrast-enhanced CT data. The models showed remarkable accuracy in predicting serosal involvement despite limitations like a retrospective single-center investigation, highlighting the possibility of clinical utility [5].

2.5 *Bayesian Network*

An adjuvant therapy screening model for advanced gallbladder cancer was created using a Bayesian network prediction model. The model showed promising accuracy in predicting survival outcomes despite obstacles, including incomplete information and biases, providing specifications for treatment options. However, there were certain limitations, such as missing factors, biases resulting from unmeasured reasons for performing RT or chemotherapy, and not being able to obtain information from the SEER dataset regarding chemotherapy specifics, radiation dosage, fields of radiation, and margin status [25] (Table 1).

All of the models used to diagnose gallbladder disease have the same basic limitation; they are all based on homogeneous and small datasets, which makes them difficult to generalize because gallbladder cancer is a rare disease, and the studies are conducted at single centers. Federated learning or a larger study would tackle this by combining data from numerous sources, increasing the amount and diversity of the dataset, and, as a result, strengthening the generalizability and robustness of the model.

3 Conventional Privacy-Preserving Methods in CSV Datasets

Methods to maintain the privacy of CSV datasets involve a number of techniques to secure sensitive information while facilitating the effective analysis of data. These techniques have the objective of ensuring individual privacy while users are interacting with huge-scale CSV datasets. In short, processes that minimize the threat of private data being misused or used by unauthorized parties comprise privacy preservers for datasets. Let's now analyze these specific privacy preservation methods used in research for CSV datasets:

One of the methods mentioned in Hidayet et al. [27] for facilitating computations on cipher data as plain text is homomorphic cryptography. Homomorphic encryption is an algorithm that can be applied to handle important data privately, such as

Table 1 State-of-the-art technologies for monitoring activity in gallbladder cancer

Paper	Dataset	Model	Evaluation metrics	Limitations
Paper [15]	GBC and Clinical Data	3D-DenseNet	AUC: 0.926	Few research subjects, missing values in clinical and treatment data
Paper [7]	US Data	Deep Learning	ACC: 74.4	Single-center data, small subgroups
Paper [13]	Gallbladder Cancer LDCT	GAN	Precision: 95.2	Not mentioned
Paper [8]	Gallbladder Cancer USG	RadFormer	ACC: 92.1	Difficulty in detecting critical features
Paper [34]	Gallbladder Data	Transfer Learning DNN	ACC: 98.35	Low contrast, aberrations
Paper [1]	Gallbladder Lesions	ResNet50, Inceptionv3	Improved specificity	Not mentioned
Paper [10]	UCF101_SBD, TRECVID	3D-DenseNet	N/A	Small dataset, lacks model output by evaluation metrics
Paper [18]	Gallbladder Lesions	CNN	AUC: 0.81, ACC: 77	Rarity of cancer, single-center study
Paper [9]	USG Images	GBCNet	ACC: 91.0	Low quality, misalignment of views
Paper [23]	CT Data	Residual CNN, ResNet50	Sensitivity, Specificity, ACC $\approx$ 98	Small patient population, rarity of disease
Paper [3]	N/A	Ensemble, ML, DL, CNN, DNN	ML using SVM: 96 67, DL: 98.77	Rapid progress in DL, small dataset, lack of details
Paper [6]	Clinical Data	Machine Learning	AUC: 0.924	Retrospective data, limited clarity due to validation in China
Paper [4]	Plasma Samples	Machine Learning	F1-score: 0.87	Smaller and realistic sample size
Paper [43]	GBC Micro-array Dataset	LASSO, SVM-RFE	AUC: 1.0	Limited samples
Paper [17]	Multi-institutional Database	CART	N/A	Retrospective dataset
Paper [14]	Serum Tumor Markers	BPNN and GA	ACC: 94.18	N/A
Paper [51]	CT Data	LASSO and RF	AUC: 0.79	Retrospective single-center study, limited generalization
Paper [5]	N/A	Radiomics	AUC: 0.94, ACC: 0.84	Single-center study on GBC serosal invasion via CT data
Paper [25]	SEER Dataset	Bayesian Network	ACC:69.67, AUC: 77.72	Incomplete variables and biases

CSV data of medical images. It makes it possible for the engines to perform some calculations on the encrypted information without having to decode the information during computation. Some of them are partially homomorphic encryption, fully homomorphic encryption, and somewhat homomorphic encryption.

Nevertheless, differential privacy provides statistical methods for preserving the private data in the database, according to Dwork [20]. The following shows how it is possible to conduct an analysis and, at the same time, reduce the likelihood of identification. Differential privacy ensures that the privacy of the data or the data query, or the result is protected by adding noise at the correct level. It permits adequate evaluation without compromising privacy, and it is vital where high standards of privacy security per user are demanded.

On the other hand, there is federated learning, which, as defined by Mohammed et al. [39], distinguishes federated learning as a kind of paradigm shift in machine learning that enables the training of models across decentralized devices or servers, though it does not have the raw data involved. It is because the model parameter is shared instead of raw data, which contributes to data protection while promoting knowledge sharing among models. FL covers issues related to data privacy, data distribution, and data ownership, and is therefore most suitable for applications of privacy-sensitive machine learning models.

In contrast, as defined in Zhao et al. [52], secure multi-party computing (SMPC) is a cryptographic process that involves data computation, so that the data can be processed among different parties without disclosing their information. Using SMPC, data computation jobs can be broken down into shared chunks performed by participants; their inputs do not have to be disclosed to compute the output, making it more secure and accurate than other forms of computation.

However, generative adversarial networks (GANs) are a promising approach for generating synthetic data that stays as close as possible to real data statistics specified in Goodfellow et al. [26]. It is a generative nature that allows a GAN to establish samples like the set of training data to use while making sure that the information given is not disclosed. The given technique is applicable to data synthesis, style transfer, and data augmentation, thus allowing one to generate privacy-preserving data.

Comparatively, mimic learning outlined in Li et al. [31] entails teaching another small neural network information from the big model it copies. Due to mimic learning, features or predictions from the bigger model are gained and transferred into the smaller model; hence, efficient training of models without compromising performance is enhanced. This approach has implications in model compression and acceleration as well as in data privacy, especially where the system is deployed in low-resource environments.

According to Sharma et al. [44], federated transfer learning is a technique in machine learning that incorporates the concepts of federated learning and transfer learning to maintain privacy. Federated transfer learning, which involves transferring learned knowledge from similar tasks or domains when training models in a decentralized manner, lets users collaborate on model building without ever having to exchange base operands. This has ramifications for private data interpretation in decentralized circumstances by using DRL models, especially in areas where data confidentiality is critical.

In the study done by Zwanenburg et al. [54], data perturbation is one of the methods that involves adding noise to data so that the purpose of the analysis can be met without infringing on an individual's privacy. This technique helps ensure that the data values or distributions are skewed in such a manner that the confidentiality of the sensitive data is maintained during the analysis process as well as when the data is being disclosed.

Lastly, similar to the work done by Swaminathan et al. [47], hash functions can be deployed frequently for content authentication owing to the fact that they do not have to invade the privacy of data due to their nature. These functions employ aspects of the data to generate hash values that can be utilized in the retrieval and identification of the data without the chances of the main details of the data being altered for any inquiries concerning them.

While traditional privacy-preserving methods are helpful for securing privacy for datasets, federated learning is the best approach. The comparison in Table 2, conceptualized following the works of Diba et al. [19] and Mistry et al. [36], reveals that, of all the methods, federated learning is the best choice. It provides while ensuring a minimal amount of data loss. Also, it protects privacy, securing the privacy of sensitive data. Federated learning is an efficient choice for purposes involving privacy preservation in datasets since it exhibits low communication complexity and moderate computational complexity [30, 37, 45, 48].

Table 2 Comparative table for CSV dataset privacy-preserving methods

Techniques	Data loss	Accuracy	Utility	Privacy	Computation	Communication
Homomorphic Encryption	High	Low	Medium	High	High	High
Differential Privacy	Medium	Medium	Medium	High	Low	Low
Federated Learning	Low	High	High	High	Medium	Low
Secure Multi-Party Computation (SMPC)	Low	High	High	High	High	High
Generative Adversarial Networks (GANs)	High	Medium	Medium	Low	High	High
Mimic Learning	Medium	High	High	High	Medium	Medium
Federated Transfer Learning	Low	High	High	High	Medium	Low
Data Perturbation	Medium	Medium	Medium	High	Low	Low
Hash Functions	High	High	High	Low	Low	Low

4 Methodology

4.1 Dataset Description

The dataset utilized in this study from Liu et al. [33] included CT scan data and clinical data from 141 patients with pathologically verified gallbladder cancer (GBC) who had undergone resection at the General Hospital Hepatobiliary Surgery Department between 2010 and 2017. The inclusion criteria were as follows: All patients confirmed a pathological diagnosis of GBC, and all had a pre-treatment contrast-enhanced CT scan. Patients with prior liver surgeries or other hepatic lesions that may obscure gallbladder lesions in studies were excluded. CT scans were performed on a 64-MDCT or 128-MDCT scanner with the same parameters used across all sites. The numerical dataset features:

- Patient identifiers are included to uniquely discover each character.
- Survival durations are recorded to track patients survival after analysis.
- Vital repute (alive or deceased) is documented to indicate the patient's survival final results at the time of the records series.
- Detailed radiomic functions encompass various aspects: Hounsfield Unit (HU) Metrics, Histogram Characteristics, Shape Parameters, and Texture Metrics.
- Key Radiomic Features Identified: GLZLM-HGZE, GLCM-Homogeneity, and NGLDM-Coarseness.

Experienced radiologists manually delineated tumor regions in the CT data that specialize in high-density, strong lesions, even in low-density regions like bile. In cases where tumor limitations had been ambiguous, radiologists recognized the precise tumor components and excluded cases of confrontation. The LIFEx software bundle facilitated the extraction of fifty-four radiomic capabilities from those segmented regions, encompassing numerous density measures (minimum, mean, standard deviation, maximum, quartiles, overall lesion glycolysis), histogram information (skewness, kurtosis, entropy, strength), shape descriptors (extent, sphericity, compactness), and texture metrics derived from the Gray Level Co-Occurrence Matrix (GLCM), Gray Level Run Length Matrix (GLRLM), Neighborhood Gray Level Different Matrix (NGLDM), and Gray Level Zone Length Matrix (GLZLM).

Dataset Preprocessing: Before version development, the dataset underwent rigorous preprocessing to cope with variations in information degrees and beautify the overall performance of the version applicable to gallbladder cancer analysis. The preprocessing commenced with normalization using a StandardScaler, making sure each feature had a median of 0 and a variance of 1. This standardization mitigates bias from differing characteristic scales, promoting the same contribution at some point in training. Additionally, polynomial functions as much as the second degree were generated by the use of polynomial features to seize nonlinear relationships, thereby increasing the feature area and doubtlessly enhancing predictive accuracy. The dataset was then broken up into test sets using train_test_split, using an 80%–20% ratio for a strong evaluation of version generalization. To keep statistics private and enhance model robustness, the training facts were partitioned into n_clients subsets, simulating disbursed datasets for federated learning. Each partition represents a consumer dataset, enabling collaborative training throughout dispensed entities without centrally aggregating raw information. For initial baseline analysis, a K-Nearest Neighbors (KNN) algorithm with n_neighbors = 5 and uniform weighting is employed, serving as a foundation to assess the FedKNN method's efficacy in predicting gallbladder cancer survival results.

4.2 Proposed Framework

Federated Learning (FL) is an approach to training models of AI on individual gadgets or servers without having to download the original data to a single location. In FL, models could be trained locally on each device, and only the modifications to the model or gradients could be sent to a central server instead of sending all the data to the central entity. According to Bonawitz et al. [11] and as shown in Fig. 1, several important factors in the decentralized training procedure are as follows:

- **Configuration:** Each of the attached devices downloads a copy of the main global model from the cloud with its help. This model is mainly prescheduled on an initial dataset and serves as a basis for training.

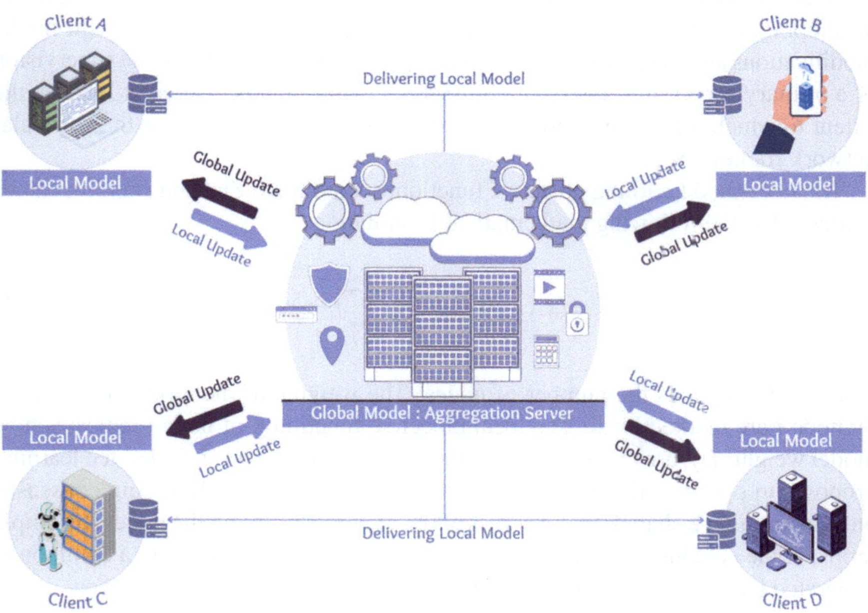

Fig. 1 Federated Learning framework with decentralized model training and central aggregation

- **Local Training:** Each device trains the model individually using its own local data, protecting user privacy, because raw data stays within the device. During this local training phase, models may adjust to the distinct aspects of the data from each device, creating customized models.
- **Secure Data Transfer:** After the local training is finished, the new model parameters or gradients are secretly transferred once more to the central server using techniques like differential privacy or encryption.
- **Aggregation:** The changes made from each contributing device are combined at the central server to create a new global model. In order to preserve confidentiality and reduce the influence of disruptive updates, sophisticated aggregation methods such as federated averaging and basic model update averaging may be used.
- **Model Deployment:** The updated global model is transmitted locally to all contributing devices to undergo additional modification through the central server following the aggregation during the model changes. Unless the model achieves the minimum level of performance or merges, this iterative method is kept going.

As stated in Kairouz et al. [28] Federated learning categories include:

Centralized Federated Learning: While utilizing centralized federated learning, all of the nodes that participate are connected to a single server, which additionally coordinates all of the different algorithmic techniques. Throughout the very first stage of training, the server is liable for identifying nodes for training and combining the model updates it is receiving. However, the server might end up turning into the system's constraint.

Decentralized Federated Learning: Nodes collaborate among themselves in developing the global model of decentralized federated learning. Although model modifications are merely distributed between affiliated nodes outside the supervision of a primary server, this approach eliminates singular points of failure. Instead, the extent to which the learning process operates might be dependent on the particular network framework.

The Federated Learning objective function, according to Zhu and Jin [53], can be expressed as the following mathematical formulation:

$$f(x_1, \ldots, x_K) = \frac{1}{K} \sum_{i=1}^{K} f_i(x_i) \tag{1}$$

Here, K specifies the number of nodes. The weights of the model as observed by node i are marked by x_i. The local objective function f_i of node i indicates how model weights (x_i) are maintained with the local dataset of node i. Federated learning proposes to optimize the objective function $f(x_1, \ldots, x_K)$ as well as gain consensus on x_i by training a shared model across all local datasets among the nodes. In simple terms, at the conclusion of the process of training, $x_1, \ldots, x_K$ merge to a common x. Federated learning algorithms that optimize the following hyperparameters:

- Total Federated Learning Iterations (T): The sum of all of the training iterations.
- Total number of nodes (K): The number of nodes that are part of the process.

- Subset of Nodes Selected for Each Round (C): The fraction of nodes utilized at each iteration.
- Local Batch Size (B): The size of the batch that is trained locally.
- Count of Iterations for Local Training Before Pooling (N): The sum of local training cycles prior to modifying the global framework.
- Local Learning Rate (η): The rate at which local model updates are developed.

FL may employ decentralized sources of data for reliable, accurate model training while sustaining privacy and minimizing the cost of communication by effectively managing these factors as stated in Mistry et al. [38]. In this study, it is proposed to apply Privacy-Preserving Federated Learning (FL) to improve gallbladder cancer (GBC) diagnosis. This is the very first time FL is being applied to GBC detection, which makes use of its decentralized layout to preserve patient privacy and maximize the collaborative capacity of multiple sources of data.

4.3 Algorithm: FedProx

FedProx modifies the local objective equation to enhance FedAvg by adding an l_2-regularization term representing the difference between the current local model and the global model learned in the previous round. Client i minimizes any disparity between its local model $\boldsymbol{\omega}_i$ and the global model as follows, as opposed to only minimizing the local loss equation φ_i on the basis of Nguyen et al. [40]:

$$\min \varphi_i + \frac{\mu}{2}\|\boldsymbol{\omega} - \boldsymbol{\omega}_i\|^2$$

Where the value of parameter $\mu \geq 0$ regulates the influence of the proximal term. The subsequent minor but crucial adjustments are made by FedProx, and these changes both enable the approach to be guaranteed to converge and produce large empirical gains in accordance with Li et al. [32]:

- Tolerating Partial Work:
  - Challenge: Diverse resource constraints impact devices, such as processing power, network connectivity, and battery life.
  - FedProx Solution: Depending on the resources available, enables varying levels of local work across devices. Instead of discarding devices that are unable to perform, it collects incomplete solutions from such devices.
  - Definition of δ_t^i-inexact Solution: A δ_t^i-inexact solution for a function $g_i(x; x^t) = G_i(x) + \frac{\lambda}{2}\|x - x^t\|^2$ is one where $\|\nabla g_i(x^*; x^t)\| \leq \delta_t^i \|\nabla g_i(x^t; x^t)\|$. A smaller δ_t^i indicates higher accuracy.

- Proximal Term:
  - Challenge: Divergence may result from an excessive number of local updates when heterogeneous information is present.
  - FedProx Solution: Adds a proximal term to the local subproblem to limit the impact of variable local updates. Instead of minimizing $G_i(x)$ alone, device i minimizes $g_i(x; x^t) = G_i(x) + \frac{\lambda}{2}\|x - x^t\|^2$.

Each step of FedProx is broken down in Algorithm 1:

Algorithm 1 FedProx

Input:
N: Total number of devices (clients).
R: Total number of communication rounds.
λ: Regularization parameter.
α_i^r: Step size for device i at round r.
w^0: Initial global model.
M: Total number of devices in the system.
p_i: Probability of selecting device i, where $i = 1, \ldots, M$.
Algorithm Steps:
for $r = 0, \ldots, R-1$ **do**
  Server selects a subset S_r of N devices at random, with probability p_i.
  Server sends the current global model w^r to all chosen devices.
  for each chosen device $i \in S_r$ **do**
    Find a α_i^r-inexact minimizer w_i^{r+1} of the local objective function:

$$w_i^{r+1} \approx \arg\min_{w} \mathcal{L}_i(w; w^r) = \mathcal{F}_i(w) + \frac{\lambda}{2}\|w - w^r\|^2$$

    Send the updated model w_i^{r+1} back to the server.
  end for
  Server aggregates in S_r to update the global model:

$$w^{r+1} = \frac{1}{N}\sum_{i \in S_r} w_i^{r+1}$$

end for

FedProx Algorithm Explanation: In a federated learning paradigm, the FedProx (Federated Proximal) algorithm defined in Algorithm 1 facilitates decentralized training of a global model by leveraging local computations on multiple edge devices. Each communication round involves the server randomly selecting a subset of N devices, each chosen with probability p_i. The server transmits the current global model, w^r, to the selected devices. Each selected device i then aims to find an α_i^r-inexact minimizer of its regularized local objective function. This local objective, $\mathcal{L}_i(w; w^r) = \mathcal{F}_i(w) + \frac{\lambda}{2}|w - w^r|^2$, incorporates a proximal term $\frac{\lambda}{2}|w - w^r|^2$ which regularizes the local model w to be close to the global model w^r. The proximal term helps to mitigate the issues arising from heterogeneous (non-iid) data distributions across devices by penalizing large deviations from the global model. Upon computing the local updates, $w^{r+1}i$, devices send these updates back to the server. The server aggregates these updates using an averaging scheme to form the new global model, $w^{r+1} = \frac{1}{N} \sum i \in S_r w_i^{r+1}$. This process iterates for a total of R communication rounds. FedProx aims to enhance the convergence properties and generalization performance of federated learning systems by addressing the statistical heterogeneity and system heterogeneity inherent in distributed data settings. The proximal term λ serves as a regularization hyperparameter, balancing the trade-off between local update fidelity and global model consistency.

5 Result and Analysis

5.1 Experimental Configuration

A computer with the specified configuration was used at the author's research lab to implement and train the models. CPU: Intel Core I7 13th Generation; RAM: 64GB; SSD: 256GB; HDD: 2TB; Graphics: RTX-4080 16 GB. In order to implement and train the models, the authors chose Python as the main programming language and popular machine learning frameworks such as Scikit-learn, TensorFlow, computer vision, neural networks, Keras, etc., as well as the Federated Learning Framework Flower.

5.2 Evaluation Metrics

A number of the most common measures that we utilized to assess the efficiency of our gallbladder cancer estimation model were Accuracy, Precision, Recall, and F1-score. The parameters that we used as variables for determining these measures include True Positives (TP), True Negatives (TN), False Positives (FP), and False Negatives (FN). For instance, TP is known to refer to correctly predicted positive cases, TN to correctly forecasted negative cases, FP to falsely predicted positive cases, and FN to falsely predicted negative cases. By employing these parameters, we intended to evaluate our model's prediction performance in an accurate way.

Accuracy: Accuracy is an evaluation of how effectively the model identifies activities generally. It refers to the proportion of correctly identified events in all cases, including true negatives and true positives. Accuracy is useful as well as simple to apply to balanced datasets. Though accuracy might appear erroneous when there is an imbalance within the datasets, Accuracy is expressed mathematically as:

$$Accuracy = \frac{True\ Positive + True\ Negative}{True\ Positive + True\ Negative + False\ Positive + False\ Negative} \quad (2)$$

Precision: Precision measures the ability of the model to distinguish the positive samples from the negative samples. Out of all the positive forecasts that are made by the model, it offers the chance of true positive estimates. If the false negative cost is less costly compared with the false positive cost, then precision is appropriate to use. Mathematically, precision is defined as:

$$Precision = \frac{True\ Positive}{True\ Positive + False\ Positive} \quad (3)$$

Recall: Recall, also referred to as sensitivity or true positive rate, refers to the ability of the model to correctly identify actual positive instances. The way to use it is by dividing the true positives by the number of false negatives added to the true positives. This measure is crucial when it is costly to ignore a positive occurrence or a false negative. Recall is expressed mathematically as

$$Recall = \frac{True\ Positive}{True\ Positive + False\ Negative} \quad (4)$$

F1-Score: F1-score is useful for datasets that are imbalanced, as it provides an option between precision and recall. This makes certain that both metrics are factored into view; this represents the harmonic average of recall and precision. If the dataset includes an unbalanced distribution of classes or it's necessary to find a balance between precision and recall, the F1-score appears useful. F1-score is expressed mathematically as

$$F1\ Score = 2 \times \frac{Precision \times Recall}{Precision + Recall} \quad (5)$$

5.3 *Performance Evaluation*

The performance assessment of the FedProx version on the training and testing sets is illustrated in Table 3. This table summarizes key evaluation metrics together with accuracy, precision, recall, and F1-score for both training and testing sets. The FedProx set of rules, using the FedKNN version, performed promising effects in both

Table 3 Evaluation metrics for FedProx model on training and testing set evaluation

Fed algorithm	Fed models	Train data				Test data			
		Accuracy	Precision	Recall	F1-Score	Accuracy	Precision	Recall	F1-Score
FedProx	FedKNN	92.86%	98.00%	94.23%	96.08%	89.66%	96.00%	92.31%	94.12%

datasets, demonstrating strong overall performance metrics suitable for gallbladder cancer analysis prediction tasks.

Analysis of Performance Metrics: The overall performance of various models, which include logistic regression and SVM, was systematically evaluated to choose the most appropriate approach for predicting survival outcomes in gallbladder cancer. Among these approaches, FedKNN applied the FedProx algorithm and tested advanced performance, reaching an accuracy of 92.86% at the training set and 89.66% on the test set. Precision, recall, and F1-score metrics always exhibited strong overall performance, with precision and recall accomplishing 98.00% and 94.23%, respectively, on the training set and keeping high values of 96.00% and 92.31%, respectively, on the testing set. These findings underscore the effectiveness of FedKNN within a federated learning framework, highlighting its capability as an innovative technique for predictive modeling within the area of gallbladder cancer analysis. FedKNN emerges as a promising technique for constructing correct predictive models in scientific studies, leveraging both radiomic functions and clinical facts successfully. Additionally, the performance metrics demonstrate a constant trend wherein rankings on the training dataset marginally surpass those on the test dataset, indicative of the model's generalization functionality.

Hyperparameter Tuning: Hyperparameters are particular parameters that cannot be learned from the data itself and need to be set by the practitioner. They are related to the learning rate, number of neurons in a neural network, or kernel size in a support vector machine. These settings regulate the learning process and the shape of the model in relation to a given task, thus striving for high accuracy. They are tuned before the training process commences and influence the model's efficiency, including accuracy and prognostic ability. Tuning can significantly improve the measures of a model's mistakes, which in some cases involve accuracy, precision, recall, and the F1-score.

In our study, hyperparameters were adjusted concerning the best performance of the federated learning model, FedProx, that we used to predict gallbladder cancer outcomes. The main parameters of the fine-tuning steps were identified in the KNN algorithm, where K is equivalent to the number of neighbors, the number of clients in the federated environment, and the number of epochs. In the case of KNN, the parameter under consideration is K, which determines the number of neighbors to consider in making a decision. This parameter is slightly different from the others as it determines how the model addresses the problem of classification by assigning the class that is most voted for by the nearest neighbors to the data point in question. We also changed values to try to find the right value that appeared to give us the highest classification rate.

As for the settings of the FedProx model, the proposed experimental studies entailed changing the number of clients involved in federated learning. This parameter determines the way data is spread and managed over different entities and may influence the outcome of the model. With reference to this parameter, the intention was to determine the optimal setup, which facilitates the use of distributed data. Further, to reduce the training time and test the validity of our methods, we tried using different numbers of epochs, where a complete pass through the training data set was done. What we did in this lab was try to find a balance in the number of epochs so that we do not over-fit the model and, on the same note, do not under-fit it either. Equilibrium or tuning is important for attaining high levels of accuracy and generalization on the training and testing sets.

As for the hyperparameters, in our implementation, we tried several combinations of these values. For example, some experiments done included K = 3, number of epochs = 10, number of clients = 2; K = 7, number of epochs = 30, number of clients = 4; and K = 5, number of epochs = 20, number of clients = 3. Finally, in our experiment, with comprehensive testing and performance comparison, we noted that when K = 5, the number of epochs is 20, and the number of clients is 3, such configurations could achieve the best performance. This setup produced reasonably good performance with high values of accuracy, precision, recall, and F1-score for both the training and test data.

Confusion Matrix Analysis: A confusion matrix, or error matrix, is another popular necessity in the field of machine learning used to determine the worth of classifier models. It provides a clear and general outlook on the model's predictions involving real label values and the different types of mistakes, including true positive, true negative, false positive, and false negative. It is very useful to check the performance and accuracy of the classification model being used.

As it can be seen from Fig. 2, the confusion matrix for the training set showed the performance of the model on the calculated training data. The matrix reveals that the model's performance was impressive, with a recall of 94.23% and precision equal to 98.00%. That is to say, the training set confusion matrix gave an understanding of having 6 true negatives, 2 false positives, 6 false negatives, and 98 true positives. This pattern gives an indication of the sort of bias the model has toward the positive class, which may be as a result of having data that was selectively sorted or as a result of the sort of classification problem under study.

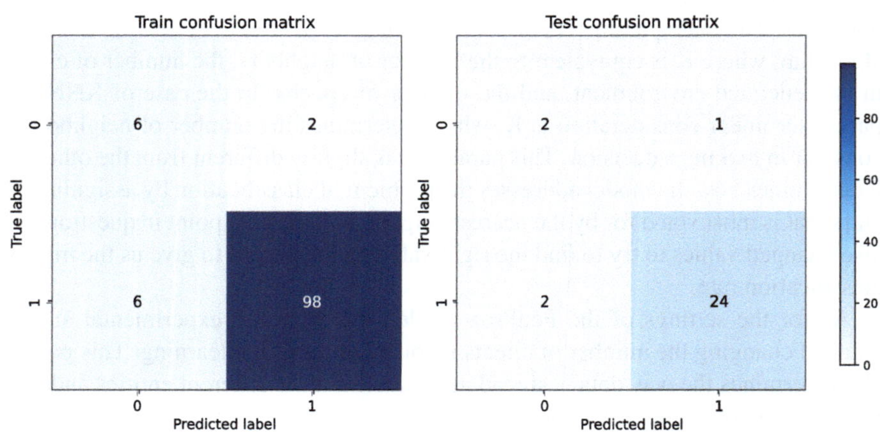

Fig. 2 Confusion matrices for the training set and test set

Likewise, the confusion matrix of the test set is as follows, and the pattern is consistent with a recall of 92.31% and precision 96.00%. The overall accuracy estimate of the model on the test set is average, with a score of about 89.66%. There are 2 false negatives, 1 false positives, 24 true positives, and 2 true negatives. The similar percentages of high recall and moderate precision when operating on the training and test sets can be interpreted as the model's high ability to identify positive cases while also suggesting a high number of false positives.

In summary, the difference in the matrices of confusion for both the training and test sets showed that the model has a preponderance of directing the outcome toward the positive class. This is so even in cases where the model demonstrated a very high recall rate compared to the precision, which is somewhat lower, especially in the test set. This tends to mean that the model, no matter how effective in flagging positive instances, will return many true positives. But if the classification is still unbalanced, it may be necessary to apply methods like resampling, changing the weights of classes, or enriching the feature set.

Computational Complexity Analysis: In this paper, we therefore present an analysis of the computational costs implicated by our proposed FL for the prognosis of gallbladder cancer. Time and space complexities are other computational resources that measure how well an algorithm keeps up with the size of the data. At the core of our work is the problem, which arises from the distributed nature of various devices, of how to perform such computations in a coordinated and efficient manner. The computational experiments in our study indicated high processing times, including a total CPU time of 3 h, 9 min, and 8 s and a total wall time of 1 h, 11 min, and 16 s. Our approach incorporates, as with the FedProx algorithm, an additional local optimization with a proximal term, thus requiring higher computations depending on the optimization technique and the proxy term. One of the peculiarities of our system is that it works on the user's devices, in contrast to the transmission of data to distant servers. Such a design helps achieve real-time prediction as well as maintain the desired device performance.

6 Conclusion and Future Work

In conclusion, this research aimed at examining the potential of federated learning, employing the FedProx algorithm as an example, in predicting outcomes in the context of gallbladder cancer prognosis based on patients' radiomic characteristics derived from CT scans and clinical information. The study proved the ability of the FedProx to improve the FedKNN model by showing improved metrics of accuracy, precision, recall, and the proposed F1-score on the training and testing data sets. Using the concept of federated learning, the described approach solves one of the crucial tasks in healthcare, which is data privacy and model generalization across various datasets. The combination of radiomic measurements and clinical features of the patient again points to the possibilities of the proposed model in the context of oncology and the value that the latter carries for personalizing medicine. As it stands, several directions for future research can be derived from this study. First, increasing the number of positive and negative cases that are available in the data could enhance the model's accuracy, particularly in the prognostication of the negative outcomes. Second, the assessment of other radiomic features and clinical characteristics may improve the model's accuracy. Third, applying validation to larger and multicenter datasets will allow for determining the model's generalizability in different patient and healthcare contexts. Furthermore, exploring further the methods for federated learning, like differential privacy, will help to secure data more and more while using the model more and more effectively. Large-scale implementation of the proposed model within clinical settings will require collaboration with healthcare institutions; only such processes will allow for assessing the practical applicability of the introduced approach and its impact on the patient's outcomes and treatment decisions in gallbladder cancer cases.

References

1. M. Akane, K. Takahashi, E. Ozawa, T. Shimakura, H. Miyaaki, K. Nakao, Recent advances in endoscopic ultrasound for gallbladder disease diagnosis. Diagnostics **14**, 374 (2024)
2. American Cancer Society, 2024. Cancer facts and figures 2024. American Cancer Society
3. A.M. Obaid, A. Turki, H. Bellaaj, M. Ksantini, Diagnosis of gallbladder disease using artificial intelligence: a comparative study. Int. J. Comput. Intell. Syst. **17**, 1–19 (2024)
4. G. Ashok, N.M. Salem, K.M. Jack, H. Gu, M. Kumar, P. Yang, V. Dinu, Machine and deep learning identified metabolites and clinical features associated with gallstone disease. Comput. Methods Progr. Biomed. Update **3**, 100106 (2023)
5. W. Bai, S. Zhou, S. Han, X. Chen, W. Pan, X. Han, X. He, Radiomics-based machine learning and deep learning to predict serosal involvement in gallbladder cancer. Abdom. Radiol. **49**, 3–10 (2023)
6. T. Bao, Y. Fu, Z. Zhong, Z. Gao, T. Ye, C. Zhang, H. Jing, Z. Xiao, Machine learning-based diagnostic model for preoperative differentiation between xanthogranulomatous cholecystitis and gallbladder carcinoma: a multicenter retrospective cohort study. Front. Oncol. **14**, 1355927 (2024)
7. G. Basu, S. Pankaj, P. Rana, U. Dutta, R. Soundararajan, D. Kalage, M. Chhabra, S. Singh, T.D. Yadav, V. Gupta, et al., Deep-learning enabled ultrasound based detection of gallbladder cancer in northern India: a prospective diagnostic study. The Lancet Regional Health-Southeast Asia (2023)
8. S. Basu, M. Gupta, P. Rana, P. Gupta, C. Arora, Radformer: Transformers with global-local attention for interpretable and accurate gallbladder cancer detection. Med. Image Anal. **83**, 102676 (2022)
9. S. Basu, M. Gupta, P. Rana, P. Gupta, C. Arora, Surpassing the human accuracy: detecting gallbladder cancer from usg images with curriculum learning, pp. 20886–20896 (2022)
10. L. Bo, X. Zhao, Q. Liu, J. Huang, L. Yu, Effective combination of 3d-densenet's artificial intelligence technology and gallbladder cancer diagnosis model. Front. Comput. Intell. Syst. **6**, 81–84 (2024)
11. K. Bonawitz, H. Eichner, W. Grieskamp, D. Huba, A. Ingerman, V. Ivanov, C. Kiddon, J. Konečný, S. Mazzocchi, B. McMahan et al., Towards federated learning at scale: system design. Proc. Mach. Learn. Syst. **1**, 374–388 (2019)
12. D. Cai, W. Xiang, et al., Privacy protection and secondary use of health data: strategies and methods. BioMed Research International (2021)

13. Y. Chang, Q. Wu, L. Chi, H. Huo, Ct manifestations of gallbladder carcinoma based on neural network. Neural Comput. Appl. **35**, 2039–2044 (2022)
14. Y. Chang, Q. Wu, L. Chi, H. Huo, Q. Li, Adoption of combined detection technology of tumor markers via deep learning algorithm in diagnosis and prognosis of gallbladder carcinoma. J. Supercomput. **78**, 3955–3975 (2021)
15. Z. Chen, T. Yin, Y. Shu, Q. Li, Z. Yuan, Y. Zhang, X. Xu, Y. Liu, A gallbladder cancer survival prediction model based on multimodal fusion analysis. Dig. Dis. Sci. **68**, 1762–1776 (2022)
16. of Clinical Oncology, A.S., et al., 2018. Gallbladder cancer: Statistics
17. G. Cotter, E.W. Beal, G.A. Poultsides, K. Idrees, R.C. Fields, S.M. Weber, C.R. Scoggins, P. Shen, C. Wolfgang, S.K. Maithel et al., Using machine learning to preoperatively stratify prognosis among patients with gallbladder cancer: a multi-institutional analysis. HPB **24**, 1980–1988 (2022)
18. S. Derya, Y. Yin, J.J. Yakar, F.J. Hoogwater, T.C. Kwee, R.J. de Haas, The value of deep learning in gallbladder lesion characterization. Diagnostics **13**, 704 (2023)
19. B.S. Diba, J.D. Plabon, M.M. Rahman, D. Mistry, A.K. Saha, M. Mridha, Explainable federated learning for privacy-preserving bangla sign language detection. Eng. Appl. Artif. Intell. **134**, 108657 (2024)
20. C. Dwork, Differential privacy, in: International colloquium on automata, languages, and programming (2006), pp. 1–12
21. C. Fitzmaurice, C. Allen, R.M. Barber, L. Barregard, Z.A. Bhutta, H. Brenner, D.J. Dicker, O. Chimed-Orchir, R. Dandona, L. Dandona et al., Global, regional, and national cancer incidence, mortality, years of life lost, years lived with disability, and disability-adjusted life-years for 32 cancer groups, 1990 to 2015: a systematic analysis for the global burden of disease study. JAMA Oncol. **3**, 524–548 (2017)
22. F. Freddie, J. Bray, I. Soerjomataram, R.L. Siegel, L.A. Torre, A. Jemal. Global cancer statistics 2018: Globocan estimates of incidence and mortality worldwide for 36 cancers in 185 countries. CA: a Cancer J Clin **68**, 394–424 (2018)
23. H. Fujita, T. Wakiya, K. Ishido, N. Kimura, H. Nagase, T. Kanda, M. Matsuzaka, Y. Sasaki, K. Hakamada, Differential diagnoses of gallbladder tumors using ct-based deep learning. Ann. Gastroenterol. Surg. **6**, 823–832 (2022)
24. A.G. Raman, R. Kulkarni, K.V. Sarma, S. Harmon, T. Sanford, H.R. Roth, Z. Xu, J. Tetreault, D. Xu, M.G. Flores, R. Raman, et al., Federated learning improves site performance in multicenter deep learning without data sharing. J. Amer. Med. Inf. Assoc. **28**, 1259–1264 (2021)
25. Z.M. Geng, Z.Q. Cai, Z. Zhang, Z.H. Tang, F. Xue, C. Chen, D. Zhang, Q. Li, R. Zhang, W.Z. Li et al., Estimating survival benefit of adjuvant therapy based on a bayesian network prediction model in curatively resected advanced gallbladder adenocarcinoma. World J. Gastroenterol. **25**, 5655 (2019)
26. I. Goodfellow, J. Pouget-Abadie, M. Mirza, B. Xu, D. Warde-Farley, S. Ozair, A. Courville, Y. Bengio, Generative adversarial nets, in *Advances in Neural Information Processing Systems*, 27 (2014)
27. U. Hidayet, A. Acar, A.S. Aksu, M. Conti, A survey on homomorphic encryption schemes: theory and implementation. ACM Comput. Surv. (Csur) **51**, 1–35 (2018)
28. P. Kairouz, H.B. McMahan, B. Avent, A. Bellet, M. Bennis, A.N. Bhagoji, K. Bonawitz, Z. Charles, G. Cormode, R. Cummings et al., Advances and open problems in federated learning. Found. Trends ® Mach. Learn. **14**, 1–210 (2021)
29. R. Kanthan, J.L. Senger, S. Ahmed, S.C. Kanthan et al., Gallbladder cancer in the 21st century. J. Oncol. **2015** (2015)
30. M.G. Kibriya, D. Mistry, D. Mistry, M.T. Rahman, S.H. Ovi, A. Tabassum, S. Hossain, Privacy-preserving vision-based detection of pox diseases using federated learning, in *Data-Driven Clinical Decision-Making Using Deep Learning in Imaging*. (Springer Nature Singapore, Singapore, 2024), pp.105–122
31. Q. Li, S. Jin, J. Yan, Mimicking very efficient network for object detection (2017), pp. 6356–6364

32. T. Li, A.K. Sahu, M. Zaheer, M. Sanjabi, A. Talwalkar, V. Smith, Federated optimization in heterogeneous networks. Proc. Mach. Learn. Syst. 429–450 (2020)
33. Z. Liu, G. Zhu, X. Jiang, Y. Zhao, H. Zeng, J. Jing, X. Ma, Survival prediction in gallbladder cancer using ct based machine learning. Front. Oncol. **10**, 604288 (2020)
34. A. Mahdi, A. Obaid Turki, H. Bellaaj, M. Ksantini, A. AlTaee, A. Alaerjan, Detection of gallbladder disease types using deep learning: an informative medical method. Diagnostics **13**, 1744 (2023)
35. N. Mansoor, F. Ali, M. Tariq, G. Kaddoum, Federated learning for privacy preservation in smart healthcare systems: a comprehensive survey. IEEE J. Biomed. Health Inform. **27**, 778–789 (2022)
36. D. Mistry, M. Mridha, M. Safran, S. Alfarhood, A.K. Saha, D. Che, Privacy-preserving on-screen activity tracking and classification in e-learning using federated learning. IEEE Access (2023)
37. D. Mistry, J.D. Plabon, B.S. Diba, S. Mukta, M.F. Mridha, Federated learning-based architecture for personalized next emoji prediction for social media comments. IEEE Access 1–1 (2024). https://doi.org/10.1109/ACCESS.2024.3448470
38. D. Mistry, M.R. Tonmoy, M.S. Anower, A.T. Hasan, Federated transfer learning for vision-based fall detection, in *International Conference on Big Data*. (Springer, IoT and Machine Learning, 2023), pp.961–975
39. K. Mohammed, S. Adnan, J.C. Cresswell, G.W. Taylor, H.R. Tizhoosh, Federated learning and differential privacy for medical image analysis. Sci. Rep. **12**, 1953 (2022)
40. H. Nguyen, L. Phan, H. Warrier, Y. Gupta, Federated learning for non-iid data via client variance reduction and adaptive server update (2022). arXiv:2207.08391
41. M. Rakić, L. Patrlj, M. Kopljar, R. Kliček, M. Kolovrat, B. Loncar, Z. Busic, Gallbladder cancer. Hepatobiliary surgery and nutrition **3**, 221 (2014)
42. P. Rawla, T. Sunkara, K.C. Thandra, A. Barsouk, Epidemiology of gallbladder cancer. Clin. Exper. Hepatol. **5**, 93–102 (2019)
43. H. Shao, J. Zhu, Y. Zhu, L. Liu, S. Zhao, Q. Kang, Y. Liu, H. Zou, Identification of characteristic genes and construction of regulatory network in gallbladder carcinoma. BMC Med. Genom. **16**, 240 (2023)
44. S. Sharma, C. Xing, Y. Liu, Y. Kang, Secure and efficient federated transfer learning (2019), pp. 2569–2576
45. M.S.H. Shovon, Z. Sultana, J. Shin, M.A. Hamid, D. Mistry, Incorporating residual connections into a multi-channel cnn for lung cancer detection in digital pathology, in *Data-Driven Clinical Decision-Making Using Deep Learning in Imaging* (Springer Nature Singapore, Singapore, 2024), p. 189
46. L.M. Stinton, E.A. Shaffer, Epidemiology of gallbladder disease: cholelithiasis and cancer. Gut Liver **6**, 172 (2012)
47. A. Swaminathan, Y. Mao, M. Wu, Robust and secure image hashing. IEEE Trans. Inf. Forensics Secur. **1**, 215–230 (2006)
48. A. Tabassum, S.H. Ovi, S. Hossain, M.R. Tonmoy, M.S.H. Shovon, M.R. Hussein, D. Mistry, Privacy preserving breast cancer prediction with mammography images using federated learning, in *Data-Driven Clinical Decision-Making Using Deep Learning in Imaging*. (Springer Nature Singapore, Singapore, 2024), pp.227–245
49. N. Truong, K. Sun, S. Wang, F. Guitton, Y. Guo, Privacy preservation in federated learning: an insightful survey from the gdpr perspective. Comput. & Secur. **110**, 102402 (2021)
50. G. Tzu-Chieh, J.A. Chao, Primary carcinoma of the gallbladder. J. Surg. Oncol. **46**, 215–221 (1991)
51. Y. Yin, D. Yakar, J.J. Slangen, F.J. Hoogwater, T.C. Kwee, R.J. de Haas, Optimal radiological gallbladder lesion characterization by combining visual assessment with ct-based radiomics. Eur. Radiol. **33**, 2725–2734 (2020)
52. C. Zhao, S. Zhao, M. Zhao, Z. Chen, C.Z. Gao, H. Li, Tan, Y.a., Secure multi-party computation: theory, practice and applications. Inf. Sci. **476**, 357–372 (2019)

53. H. Zhu, Y. Jin, Multi-objective evolutionary federated learning. IEEE Trans. Neural Netw. Learn. Syst. **31**, 1310–1322 (2019)
54. A. Zwanenburg, S. Leger, L. Agolli, K. Pilz, E.G. Troost, C. Richter, S. Löck, Assessing robustness of radiomic features by image perturbation. Sci. Rep. **9**, 614 (2019)

Cataract Detection Using Federated Learning: A Decentralized Approach for Cross-Institutional Collaboration of Data Parameters

Md. Azizul Hakim Shourov, Abu Shaleh Md. Kaium, Rubaiyat Rahman, Md. Monzurul Kabir Zhanda, and Jahir Ibna Rafiq

Abstract Over 2.2 billion people globally suffer from cataracts, which make early detection and treatment extremely difficult. The limitations of traditional diagnostic approaches arise from the absence of specialized knowledge and costly equipment. The absence of diverse and large-scale datasets impedes the progress of artificial intelligence (AI)-based cataract detection systems, which limits their accuracy to the clinical level and prioritizes patient data privacy while addressing dataset constraints by utilizing Federated Learning (FL) as a novel solution. Federated learning is a decentralized approach in machine learning that combines model updates either locally or at a central server, allowing multiple local devices to collaboratively develop a global model without the need to exchange raw data. This allows for model improvement across distant datasets while maintaining data privacy. We trained the Local Models (LM) simultaneously on multiple local servers using different datasets. Updated learning models (LMs) are combined using FL approaches to create a strong Global Model (GM), which improves detection accuracy and enables inter-institutional collaboration while preserving patient privacy. The paper focuses on putting this FL-based architecture into practice and evaluating it, providing a scalable and private cataract detection method. This research presents a scalable and privacy-preserving cataract detection method, leveraging FL to enhance diagnostic accuracy while addressing the global need for secure and effective tools in early cataract detection.

Keywords Convolutional Neural Network · Federated averaging · Federated learning · Global model · Local model

---

Md. Azizul Hakim Shourov, Abu Shaleh Md. Kaium, Rubaiyat Rahman are contributed equally to this work.

Md. A. H. Shourov · A. S. Md. Kaium · R. Rahman · Md. M. K. Zhanda
Department of Computer Science, American International University-Bangladesh, Dhaka, Bangladesh

J. I. Rafiq (✉)
Biomedical Engineering Department, University of Toronto, Toronto, Canada
e-mail: JahirIbna.Rafiq@uhn.ca

M. F. Mridha and N. Dey (eds.), *Federated Learning in Health Care Technology*, Studies in Computational Intelligence 1216, https://doi.org/10.1007/978-981-96-8353-6_10

1 Introduction

A new study released by the World Health Organization indicated that the global prevalence of vision impairment is 2.2 billion as of 2023. Cataracts primarily cause visual impairment or blindness and affect approximately 94 million individuals. Vision impairment is a significant global economic burden, with the yearly cost of lost productivity estimated at US$ 411 billion.

Conventional cataract detection methods involve technology such as slit lamps and fundus cameras, which incur substantial costs and require specialization for diagnostic operations. A lack of qualified ophthalmologists poses a significant challenge. This results in an extension of cataract diagnosis, necessitating medical intervention [10]. Recently, cataract detection methods that leverage machine learning and artificial intelligence models have attracted scientific interest [7]. This demonstrates the novel opportunities and progress in the field of ophthalmology for cataract detection.

Sophisticated machine learning and deep learning models contain millions of parameters that must be acquired from adequately extensive, curated datasets for clinical-grade accuracy [4]. Collaborative learning, which uses multi-institutional data to train a singular model, is an effective method for augmenting both the data size and variety [4]. However, acquiring such data is challenging because of the sensitivity of health information and stringent regulations governing its use [3]. While data anonymization can address some of these limitations, it is widely acknowledged that simply removing metadata, such as patient names or birth dates, is not enough to ensure privacy protection [21]. This not only imposes constraints on cross-institutional data sharing but also restricts the understanding of the worldwide state of the disease.

To address these issues, we propose an architecture based on Federated Learning (FL) to safeguard the confidentiality of patients' medical data and to facilitate the classification of cataract detection scenarios in collaboration with numerous medical institutions. This document primarily addresses the implementation and evaluation of our architecture, achieving superior accuracy compared with other contemporary models.

Machine Learning models were concurrently trained using datasets of fundus images across various local servers. Upon completion, revised Local Models (LM) are transmitted for aggregation via a FL approach to a Global Server to create a Global Model [4]. This method updates both LM and GM. This facilitates inter-institutional collaboration and guarantees that sensitive data remain decentralized.

The present condition of cross-institutional interactions in the medical field necessitates institutions to consolidate patient data in a centralized repository for model training [4]. A data sample was gathered from various hospitals and transmitted to initialize a local model. Differential privacy is used as an adjunct to enhance data encryption to achieve security objectives. The parameters were subsequently transmitted to the cloud server for the aggregate. The cloud server returns the weighted average, resulting in an update to the local model. The data remained decentralized throughout the procedure.

This concept seeks to create a decentralized foundation for a privately focused data cooperation system. Conventional deep learning models have demonstrated substantial accuracy through the application of weighted averages. Federated Inception RestNet V2, FedCNN, and FedInceptionV3. Federated Inception RestNet V2 demonstrated the highest accuracy, achieving a result of 98%. This result was obtained using 2000 data samples.

In summary, the key contributions of this study are outlined as follows:

1. We aim to develop an optimal methodology for precise classification and detection of cataracts.
2. We propose a robust approach using a challenging dataset derived from a Kaggle competition and validate the effectiveness of our model through meticulous analysis and experimentation.
3. We adopted machine learning techniques to classify and detect cataracts based on clinical parameters, utilizing the Federated Averaging algorithm for its superior performance in FL.
4. We integrated differential privacy mechanisms to enhance data encryption and ensure robust security and data protection.

The structure of the paper is outlined as follows: Sect. 2 offers a detailed review of prior research in this field Sect. 3 describes the proposed methodology. Section 4 details the framework explanation. Section 5 examines the experiment and analyzes the result. Section 6 discusses the limitations and future research directions and concludes the study.

2 Literature Review

Advancements in decentralized machine learning methodologies have evolved markedly, especially within the medical domain. Sheller et al. [23] emphasized the promise of FL for collaborative medical data analysis, specifically in brain tumor segmentation. Their research illustrated how FL facilitates several institutions in training a global model while safeguarding sensitive patient data, thereby overcoming a major barrier to inter-institutional data exchange. This decentralized approach preserved patient confidentiality while obtaining excellent precision, with the Federated-averaging algorithm achieving an average accuracy of 0.91 across many institutions. This study demonstrates the groundbreaking possibilities of FL for enhancing data collaboration throughout healthcare. Choudhury et al. [3] furthered this by incorporating Differential Privacy (DP) methods into FL to improve privacy in data sharing. Their methodology, which examined sensitive health data across institutions, attained an accuracy of 0.85 while preventing privacy breaches. Scientists observed how FL facilitated collaborative learning among organizations that were otherwise unwilling to exchange data due to regulatory and privacy issues. This has opened new prospects for the medical sector, especially for the establishment of extensive joint databases for rare conditions or circumstances characterized by data scarcity. Rieke et al. [20]

explained the issue of data heterogeneity in multi-institutional collaborations. Federated Learning presents significant potential for medical data sharing; however, disparities in dataset quality and size among institutions adversely affect model performance. A study on skin cancer diagnosis utilizing Federated Learning reported an AUC of 0.87. This study highlights the necessity of utilizing advanced aggregation methods to guarantee fair contributions from involved institutions and mitigate data imbalances in collaborative, multi-institutional contexts. Literature Review 2 Xu et al. [31] applied federated learning to analyze cardiovascular disease data, demonstrating the importance of data collaboration among five hospitals. The collaborative model presented an accuracy of 0.88 in predicting cardiovascular events. This study demonstrates that federated learning can enhance collaborative research by enabling institutions to collectively improve their models while maintaining patient confidentiality. They observed that enhanced security measures were required to safeguard collaborative processes against adversarial attacks.

Zhang et al. [33] applied conventional machine learning methods for cataract classification while highlighting the risks associated with centralized data storage, particularly in sensitive medical domains. They proposed that Federated Learning could transform data collaboration in ophthalmology by decentralizing the data while preserving privacy. Their CNN model, developed using a centralized dataset, achieved an accuracy of 0.95. However, subsequent efforts may concentrate on implementing federated learning for enhanced security and extensive data collaboration in this field. Imran et al. [5] utilized semi-supervised learning for cataract identification, leveraging datasets from three hospitals, and achieved a high accuracy of 0.96. They encountered challenges in data sharing due to privacy concerns, leading to the suggestion of Federated Learning (FL) as a solution. FL permits numerous hospitals to jointly train models while maintaining decentralized patient data, thereby facilitating extensive collaboration among healthcare facilities and enhancing diagnostic precision through collective learning. Gabrielli et al. [4] investigated the influence of FL on improving the diagnostic precision in ophthalmology, specifically in the identification of cataracts. Their review emphasized that federated learning allows institutions to collaborate without disclosing raw data, attaining an accuracy similar to that of centralized approaches. They highlighted the capacity of federated learning to significantly enhance data collaboration in the medical field, where data privacy is paramount, particularly in ophthalmic diagnoses.

Moon and Lee [14] conducted research on privacy-preserving methodologies in the context of data sharing in medical artificial intelligence. They established a Federated Learning (FL) architecture that integrated homomorphic encryption and differential privacy, allowing institutions to share model updates while protecting sensitive medical information. The methodology was effectively implemented across many medical datasets, attaining an accuracy of 0.88 in diagnosing retinal disorders, while maintaining stringent data protection throughout the collaborative process. Nizami and Gulani [17] investigated secure data collaboration in healthcare by utilizing federated learning alongside secure multiparty computation. Their program achieved an accuracy of 0.90 in categorizing cataract severity while maintaining privacy using sophisticated encryption methods. The authors emphasized the usefulness of federated learning in enabling extensive medical research collaborations without requiring the centralization of patient data, a vital advantage in ophthalmology where patient data confidentiality is essential.

3 Methodology

This section describes the crucial steps involved in acquiring and preparing the dataset for the proposed framework. First, the sources were described: fundus pictures from three hospitals that provided samples of both cataract and non-cataract patients. Initial processing techniques, including image enhancement, noise reduction, and standardization, are applied to ensure the accuracy and quality of the dataset. Some difficulties that occurred during the data collection process are also discussed. However, solutions have been outlined to overcome these difficulties. Thus, a reliable model training and accurate cataract detection method were developed using this comprehensive approach.

3.1 Dataset Preparation

In this study, two types of fundus images, cataract and non-cataract, were collected from three hospitals. Datasets with approximately 200 images were processed in crucial steps to verify their suitability for model training.

3.1.1 Image Enhancement

he images were enhanced to improve quality. This step includes adjusting the contrast and brightness to make the cataracts more detectable.

3.1.2 Noise Reduction

Noise reduction techniques have been applied to mitigate the effects of unnecessary artifacts. This process removes unwanted noise from the images, helping to obtain clearer data for accurate model training.

3.1.3 Normalization

The pixel values were normalized to standardize the input data, aligning them between the range 0 and 1, the consistency of the data, fed into the model was confirmed.

3.2 Machine Learning Models

FedCNN

The Federated Convolutional Network (FedCNN) is a convolutional neural network that is adopted for federated learning. Moreover, it can be designed to detect cataracts in fundus images while simultaneously, preserving privacy at the same time.

- **Architecture**
  - **Convolutional layers:** FedCNN employs several convolutional layers to detect local features in fundus images. These convolutional layers are formed by activation functions (ReLU) and pooling layers to mitigate the dimensions and capture the essential features.
  - **Fully connected layers:** The final classification is achieved by utilizing the fully connected layers based on the extracted features.
- **Training process**
  - **Local training:** Each local node is trained using FedCNN itself on subset of the data. This process is accomplished based on the local training data through computer gradients and by updating the model parameters.
  - To update the global model, the parameters of the local model are aggregated using Federated Averaging at the central server.

FedInceptionV3

FedInceptionV3 is a convolutional neural network optimized for federated learning that relies on the InceptionV3 architecture. Moreover, it can perform advanced picture classification tasks, such as cataract identification.

- **Architecture**
  - **Inception modules:** This model utilizes inception modules that integrate multiple filter sizes inside a single layer. This design allows the network to capture the features at various scales.
  - **Global average pooling:** FedInceptionV3 decreases the dimensionality of the feature maps prior to classification by employing global average pooling in place of fully connected layers.
- **Training process**
  - **Local training:** Each federated node trains its FedInceptionV3 model using local data. The model parameters are updated at a local level whereas the gradients are transmitted to a central server.
  - **Aggregation:** The model parameters are accumulated by the central server model from all nodes to update the worldwide FedInceptionV3 model.

FedInceptionResNetV2
FedInceptionResNetV2 is a cutting-edge model that combines the advantages of the InceptionV3 and ResNet architectures. This model was optimized for federated learning to enhance cataract detection capabilities.

- **Architecture**
  - **Inception modules with residual connections:** Through residual connections, FedInceptionResNetV2 integrates inception modules. Both the residual connections and inception modules improve feature extraction, which is a common issue in deep networks produced by vanishing gradients.
  - **Deep network structure:** The deep architecture of the model permits the development of complex features from fundus images more effectively.
- **Training process**
  - **Local training:** Every node uses its own data subset to train its FedInception-ResNetV2 model. The computation of the gradients is communicated through the central server.
  - **Aggregation:** Parameters from the local models are aggregated at the central server to modify the global model.

3.3 Differential Privacy Integration

Incorporating differential privacy techniques into a federated learning framework enhances data security during the training process. Differential privacy adds noise to gradients to secure individual data points while contributing to global model training.

3.4 Proposed Methodology

This section offers a detailed overview of the dataset. Initially, the operational process of the proposed framework is described. Then, we illustrate how federated learning with differentiated privacy protects user privacy.

Traditional techniques for identifying cataracts in eye health evaluations typically require expensive and difficult hospital visits. Conventional machine learning algorithms can assist in detecting cataracts in images. However, this method raises concerns about the privacy of user data. The proposed technique prioritizes user privacy and security while addressing the pressing demands for cataract diagnosis through Federated Learning. Our framework uses the FedCNN, FedInceptionV3, and FedInceptionResNetV2 architectures to classify cataracts. Differential privacy is employed to safeguard data privacy when local model parameters collaborate with global models while maintaining model validity. Figure 1 shows the result of the

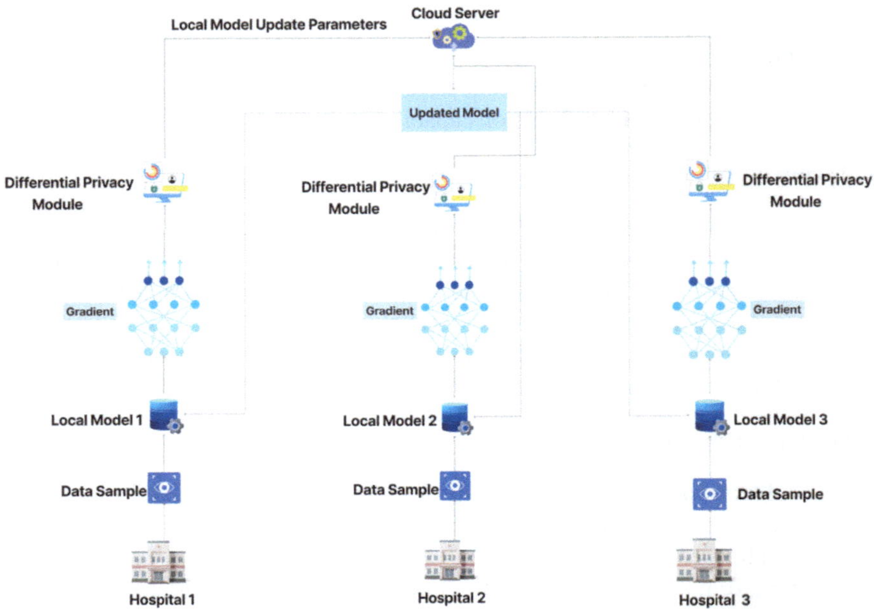

Fig. 1 Data samples from hospitals are encrypted with differential privacy before initializing a local model, then sent to a cloud server for aggregation. The server computes a weighted average to update the local model, ensuring decentralized data privacy and security

proposed design. First, a detailed representation of the dataset was provided. The manner in which the proposed framework works sequentially is explained bellow. Finally, the privacy protection feature of federated learning was demonstrated.

4 Framework Explanation

The proposed algorithm identifies cataracts by using a federated learning method that prioritizes user privacy. This study prioritizes user privacy and uses a federated learning technique to build a robust model for cataract identification. Data were collected from three hospitals, each with two types of fundus image: cataract and non-cataract. The dataset was split into two parts: 80% for training and 20% for testing the local training. To ensure correct evaluation, three frameworks were applied for training the local models: FedCNN, FedInceptionV3, and FedInceptionResNetV2. The objective of identifying the most accurate design for cataract identification. After local training, the parameters were forwarded to a global model to avoid data concentration while safeguarding user privacy. Differential privacy approaches are used to secure privacy while transferring parameters. This strategy secures user data from illegal access while maintaining model accuracy and reliability by safeguarding the confidentiality of local data inputs. Federated averaging enables local models to train the global model and strengthen collaborative learning while maintaining data privacy. The proposed technique improves traditional machine learning approaches by decentralizing training and using differential privacy. Moreover, this technique protects sensitive medical data during training. However, the proposed technique enhances the security and privacy of sensitive medical data while maintaining the accuracy of the model, thereby providing a reliable foundation for rapid cataract detection. It prioritizes privacy through federated learning, making it ideal for medical applications that require data security.

Algorithm 1: Federated learning algorithm with FedAvg and Differential Privacy

Initialize the global model M_{global} for cataract detection.
The clients were divided into two groups: trainers and classifiers.
  For each trainer client $i = 1, \ldots, n_{\text{trainers}}$ **in parallel**
    Copy the global model to the local model: $M_i \leftarrow M_{\text{global}}$
    for $e = 1$ to E_{trainers}
      Shuffle the trainer client's dataset D_{train_i}
      for each mini-batch $(x_1, y_1), \ldots, (x_B, y_B)$ in D_{train_i} do
        Calculate the gradient g: $g \leftarrow \frac{1}{B} \sum_{j=1}^{B} L(M_i(x_j), y_j)$ for the
        cataract detection task.
        Add noise to the gradient for **differential privacy**: $g \leftarrow g +$ noise
        Update the local model M_i using Stochastic Gradient
        Descent(SGD): $M_i \leftarrow M_i - \eta \times g$
      end for (mini-batch loop)
    end for (local epoch loop)
  end for (trainer client loop)
  Aggregate models using Federated Averaging (FedAvg):
  $M_{\text{global}} \leftarrow \frac{1}{n_{\text{trainers}}} \sum_{i=1}^{n_{\text{trainers}}} M_i$
  for each classifier client $j = 1, \ldots, n_{\text{classifiers}}$ in parallel
    Use the global model M_{global} to classify the local dataset D_{classify_j}
  end for (classifier client loop)
end for (communication round loop)

Output: Trained global model M_{global} for cataract detection.

4.1 Algorithm Explanation

The algorithm begins by initializing a global model, denoted as M_{global}, which is tailored for cataract detection. Following this initialization, clients are partitioned into two distinct groups: trainers and classifiers. The training phase commences by iterating each trainer client in parallel. For each trainer client, a local model, denoted as s M_i, is instantiated as a copy of the global model M_{global}. Subsequently, a predetermined number of training epochs and E_{trainers} are executed. Within each epoch, the trainer client dataset (D^i_{train}) was shuffled, and training was performed via mini-batch processing. The gradient (g) was computed with respect to the loss function (L) for cataract detection. This gradient guides the model update process through stochastic gradient descent (SGD), where the local model M_i is adjusted using the learning rate η.

In the aggregation step, federated averaging was used to combine the models from all training clients. This approach involves averaging the model parameters among trainer clients, providing an updated global model, M_{global}. Subsequently, in the classification phase, each classifier client utilizes the updated global model M_{global}, to perform classification tasks on its local dataset D^j_{classify}. This cyclical training procedure, the processes of aggregation and classification remains ongoing for a certain number of communication rounds, represented as T. Finally, the method generates the trained global model, M_{global}, designed for the detection of cataracts.

The equations utilized in this method are essential for directing training and aggregation processes. The gradient calculation equation determines the gradient (g) by calculating the mean of the gradients of the individual data points (x_j, y_j) in relation to the loss function (L). Subsequently, in the model update equation, the local model parameters (M_i) are updated using Stochastic Gradient Descent (SGD), wherein the gradient (g) is modulated by the learning rate (η). Ultimately, the federated averaging equation calculates the updated global model (M_{global}) by averaging the parameters across all trainer clients, thereby providing collaborative learning while preserving privacy.

Equations used:

$$\text{Gradient Calculation: } g = \frac{1}{B}\sum_{j=1}^{B} L(M_i(x_j), y_j) \tag{1}$$

$$\text{Model Update: } M_i = M_i - \eta \times g \tag{2}$$

$$\text{Federated Averaging: } M_{\text{global}} = \frac{1}{n_{\text{trainers}}}\sum_{i=1}^{n_{\text{trainers}}} M_i \tag{3}$$

In the gradient calculation in Eq. (1), the gradient (g) is determined as the mean of the gradients of the individual data points (x_j, y_j) about the loss function (L). This computation is crucial for determining the direction and degree of the necessary modifications. The model parameters were changed throughout the training phase. Each piece of information adds to the overall gradient, which affects the learning process of the model.

The model update method in Eq. (2) depends on the stochastic gradient descent (SGD). The local model parameters (M_i) were modified by removing the product of the learning rate (η) and gradient (g). This update process ensures that the model advances toward reducing the size of the loss function, thereby improving its performance in cataract detection.

Finally, Eq. (3) implements the federated averaging technique for updates to the global model (M_{global}). The modified global model was calculated as the mean of the parameters for all trainer clients (n_{trainers}). This collaborative approach enables knowledge sharing among clients while preserving their privacy, as each client contributes to the collective improvement of the global model without sharing their individual data.

5 Experiment and Result Analysis

To obtain the best outcome, we conducted different tests with different deep learning models such as CNN, InceptionV3, and InceptionResNetV2 to classify two different types of data. The dataset used in this experiment contained approximately 2000 fundus images of cataracts and non-cataracts.

5.1 Dataset

In this study, more than 2000 fundus images were used, and each image was labeled and divided into two categories. 1038 and 1074 for cataracts and 1074 for non-cataracts, respectively. Table 1 highlights the distributed number of datasets.

5.2 Experimental Setup

A source editor, Visual Studio code was used to run all the necessary code for the paper.The models were developed using the Python programming language, utilizing well-known machine learning frameworks like TensorFlow, Keras, and Scikit-learn. Various extensions have been used to efficiently run all source codes.

Table 1 Dataset table

Types of class	Cataract	Normal
Size	1038	1074

5.3 Evaluation Matrices

Four performance metrics were used to observe the effectiveness of the three pre-trained deep learning models in a federated environment. To test these models, the data were randomly divided into 80% and 20% for training and testing, respectively. The proposed model was evaluated using the following equation.

- Accuracy: Measures overall correctness.

$$Accuracy = \frac{(TruePositive + TrueNegative)}{TotalNumberOfPredictions}$$

- Precision: Evaluate positive prediction quality.

$$Precision = \frac{TruePositive}{(TruePositive + FalsePositive)}$$

- Recall: Assesses sensitivity to positive instances.

$$Recall = \frac{TruePositive}{(TruePositive + FalseNegative)}$$

- F1-Score: Balances Precision and Recall

$$F1 - Score = \frac{2 * (precision * recall)}{(precision + recall)}$$

To fully analyze the effectiveness of classification models, these evaluation metrics are considered.

5.4 Performance Evaluation

In this study, a dataset of more than 2000 images split into two classes was used to assess the performance of three different federate for patient cataract identification. Several criteria are used to assess the performance: in Sect. 5.4.1, performance metrics are analyzed; in Sect. 5.4.3, confusion matrix analysis concludes the discussion.

5.4.1 Performance Metrics Analysis

Tables 2 and 3 the performance metrics of the three federated learning models. FedInceptionCNN with 77.22%, FedInceptionV3 with 94.77%, and FedInceptionResNetV2 with 96.91% performed the best overall in federated averaging training.

Table 2 Performance metrics of the federated models on the training data

Federated model	Accuracy	Precision	Recall	F1-score
FedCNN	0.77	0.77	0.92	0.84
FedInceptionV3	0.94	0.95	0.98	0.96
FedInceptionResNetV2 (Proposed)	0.96	1.00	0.96	0.98

Table 3 Performance metrics of the federated models on the test data

Federated model	Accuracy	Precision	Recall	F1-score
FedCNN	0.89	0.89	1.00	0.94
FedInceptionV3	0.90	0.90	0.98	0.93
FedInceptionResNetV2 (Proposed)	0.98	0.97	1.00	0.99

We used the macro-precision matrix, which was computed by averaging the accuracy scores of all local models, as the precision measurement. The unweighted average of these scores is then determined by macro-precision, which computes the accuracy separately for each local model [53]. Similar trade-offs can be seen here, as all models demonstrated good results, with the FedCNN value being 77% and the FedInceptionResNetV2 value being 100%. FedInceptionV3, FedInceptionResNetV2, and FedCNN the highest weighted recall metrics, with values of 98%, 96%, and 92%. FedInceptionResNetV2 had the highest value of 98 for the F1-score metric, which considered both precision and recall. FedInceptionV3 followed with a score of 96%, while FedCNN came in third with 84% (Fig. 2).

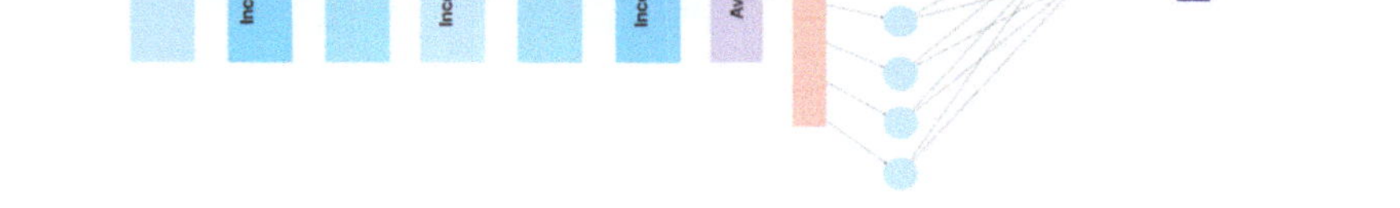

Fig. 2 FD approach based on Inception-ResNetV2 architecture, enabling collaborative model training for data sample, ensuring decentralization

5.4.2 Loss and Accuracy Analysis

In machine learning, graphs of accuracy and loss are useful tools for evaluating the model performance. The accuracy graph shows the successfully predicted values and the loss graph shows the errors between the predicted and real-time values throughout training. Machine learning problems such as overfitting and under-fitting can be observed using these graphs. Approaches for improving the model, such as hyperparameter tuning and model selection, may be guided by comparing the accuracy and loss graphs [53].

The loss and accuracy graphs for all models were generated using Matplotlib. Personalized FL models were applied to virtual devices during the simulation to manage the training and updating of the models. Fundus images of cataracts and normal image datasets of a non-iid nature were considered and the data samples are distributed differently throughout the virtual devices to simulate real-world situations.

These graphs offer important perspectives on how personalized federated learning models perform when confronted with nonID data distributions. The simulation provided useful insights into the strengths and drawbacks of FedCNN, FedInceptionV3, and FedInceptionRsNetV2 in a distributed federated learning environment via a systematic performance comparison. The results are shown in Fig. 3.

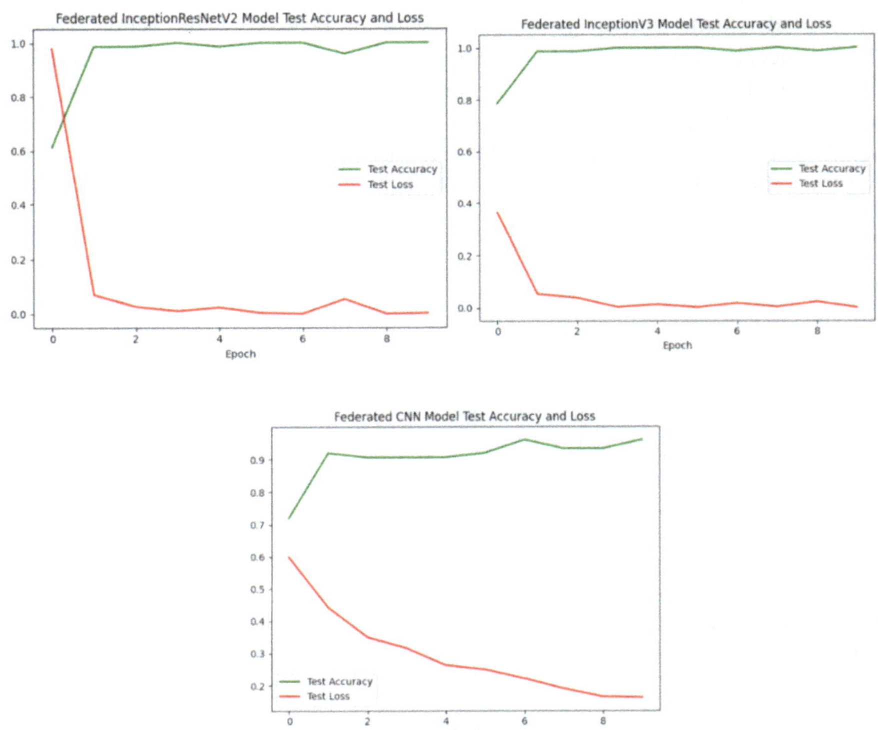

Fig. 3 Loss and accuracy graph of FedInceptionResNetV2, FedInceptionV3, and FedCNN

5.4.3 Confusion Matrix Analysis

The confusion matrix is a crucial tool for evaluating the effectiveness and accuracy of the classification model, as it highlights various types of errors made by the model, including false positives, false negatives, true positives, and true negatives [53]. Figure 6 shows the confusion matrix, which provides insight into the performance of different models using the given dataset. FedInceptionResNetV2 clearly performs the best in terms of sample classification. However, although there was a considerable difference, the other models also performed well in terms of categorization accuracy. These results demonstrate the performance of FedInceptionResNetV2 and other models in accurately identifying the experimental data, with FedCNN being the least successful model. The results are shown in Fig. 4:

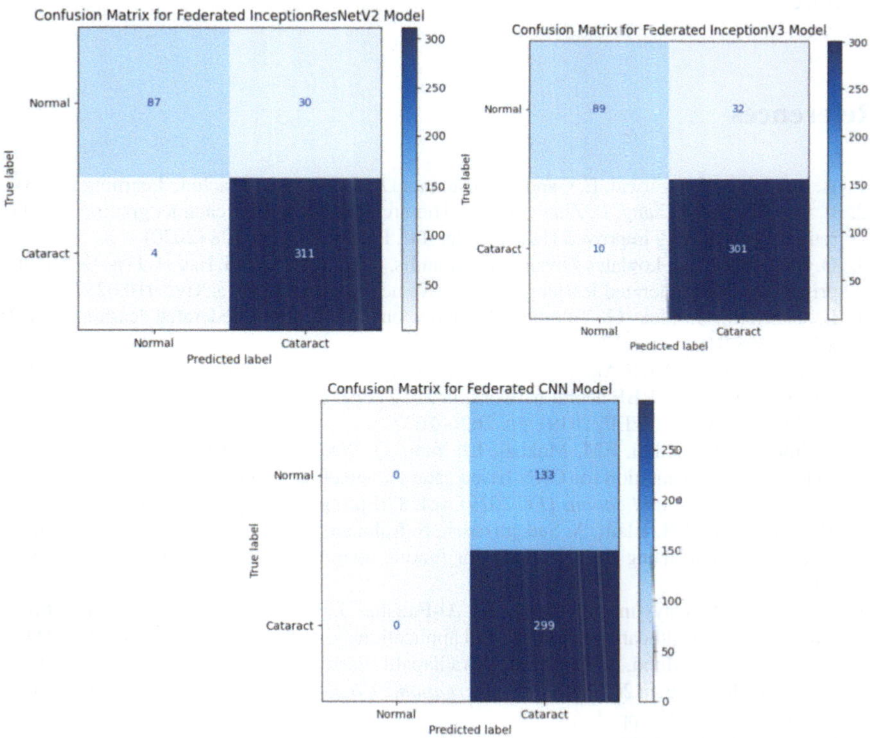

Fig. 4 Confusion matrix for FedInceptionResNetV2, FedInceptionV3, and FedCNN

6 Conclusion and Future Work

One of the most prevalent conditions that distorts vision is cataract [20]. The likelihood of blindness and degeneration can be significantly reduced by routine screening and early intervention [5]. Cataracts can be easily detected using classification methods and in this paper, we present a decentralized model for cataract detection. The numerical results demonstrated the accuracy of the proposed methodology for cataract detection. By being proactive, a model's capabilities can be improved. For example, a larger training dataset may improve the accuracy rate of the model. Cryptographic encryption can be used to improve the model resilience. In federated learning, safe communication between local and global models is ensured through the widespread use of homomorphic encryption and secure multiparty computation. Methods for grading cataract severity can be applied in conjunction with security measures.

References

1. A. Agnihotri, T. Agarwal, E. Gandotra, Cataract Detection using Machine Learning (2023)
2. L. Cao, H. Li, Y. Zhang, L. Zhang, L. Xu, Hierarchical method for cataract grading based on retinal images using improved Haar wavelet. Inf. Fusion **53**, 196–208 (2020)
3. O. Choudhury, A. Gkoulalas-Divanis, T. Salonidis, I. Sylla, Y. Park, G. Hsu, A. Das, Differential privacy-enabled federated learning for sensitive health data (2019). arXiv:1910.02578
4. E. Gabrielli, G. Pica, G. Tolomei, A survey on decentralized federated learning (2023). arXiv:2308.04604
5. A. Imran, J. Li, Y. Pei, F. Akhtar, J.J. Yang, Q. Wang, Cataract detection and grading with retinal images using SOM-RBF neural network, in *2019 IEEE Symposium Series on Computational Intelligence (SSCI)* (IEEE, 2019), pp. 2626–2632
6. A. Imran, J. Li, Y. Pei, F.M. Mokbal, J.J. Yang, Q. Wang, Enhanced intelligence using collective data augmentation for CNN based cataract detection, in *Frontier Computing: Theory, Technologies and Applications (FC 2019)*, vol. 8. (Springer Singapore, 2020), pp. 148–160
7. M.S. Junayed, M.B. Islam, A. Sadeghzadeh, S. Rahman, CataractNet: An automated cataract detection system using deep learning for fundus images. IEEE Access **9**, 128799–128808 (2021)
8. N. Khalid, A. Qayyum, M. Bilal, A. Al-Fuqaha, J. Qadir, Privacy-preserving artificial intelligence in healthcare: techniques and applications. Comput. Biol. Med. 106848 (2023)
9. H.B. Koncha, P. Mohd, S. Ravikanti, P. Yadlapalli, Early detection and prediction of cataract using deep learning, in *2023 IEEE 8th International Conference for Convergence in Technology (I2CT)* (IEEE, 2023), pp. 1–6
10. C.J. Lai, P.F. Pai, M. Marvin, H.H. Hung, S.H. Wang, D.N. Chen, The use of convolutional neural networks and digital camera images in cataract detection. Electronics **11**(6), 887 (2022)
11. L. Li, Y. Fan, M. Tse, K.Y. Lin, A review of applications in federated learning. Comput. & Ind. Eng. **149**, 106854 (2020)
12. G. Litjens, T. Kooi, B.E. Bejnordi, A.A.A. Setio, F. Ciompi, M. Ghafoorian, J.A. Van Der Laak, B. Van Ginneken, C.I. Sánchez, A survey on deep learning in medical image analysis. Med. Image Anal. **42**, 60–88 (2017)
13. D. Mistry, M.F. Mridha, M. Safran, S. Alfarhood, A.K. Saha, D. Che, Privacy-preserving on-screen activity tracking and classification in E-learning using federated learning. IEEE Access (2023)

14. S. Moon, W.H. Lee, Privacy-preserving federated learning in healthcare, in *2023 International Conference on Electronics, Information, and Communication (ICEIC)*. (IEEE, 2023), pp. 1–4
15. C. Muchibwa, M.H.S. Eldaw, Mu. Mu, M.O. Agyeman, An assessment of contemporary methods and data-enabled approaches for early cataract detection. F1000Research **12**, 998 (2023)
16. D.C. Nguyen, Q.V. Pham, P.N. Pathirana, M. Ding, A. Seneviratne, Z. Lin, O. Dobre, W.J. Hwang, Federated learning for smart healthcare: a survey. ACM Comput. Surv. (CSUR) **55**(3), 1–37 (2022)
17. A.A. Nizami, A.C. Gulani, Cataract. In StatPearls [Internet] (StatPearls Publishing, 2022)
18. T. Pratap, P. Kokil, Automatic cataract detection in fundus retinal images using singular value decomposition, in *2019 International Conference on Wireless Communications Signal Processing and Networking (WiSPNET)*. (IEEE, 2019), pp. 373–377
19. A. Qayyum, K. Ahmad, M.A. Ahsan, A. Al-Fuqaha, J. Qadir, Collaborative federated learning for healthcare: Multi-modal covid-19 diagnosis at the edge. IEEE Open J. Comput. Soc. **3**, 172–184 (2022)
20. N. Rieke, J. Hancox, W. Li, F. Milletari, H.R. Roth, S. Albarqouni, S. Bakas, M.N. Galtier, B.A. Landman, K. Maier-Hein, S. Ourselin, The future of digital health with federated learning. NPJ Dig. Med. **3**(1), 119 (2020)
21. L. Rocher, J.M. Hendrickx, Y.A. De Montjoye, Estimating the success of re-identifications in incomplete datasets using generative models. Nat. Commun. **10**(1), 1–9 (2019)
22. N. Rodríguez-Barroso, D. Jiménez-López, M.V. Luzón, F. Herrera, E. Martínez-Cámara, Survey on federated learning threats: Concepts, taxonomy on attacks and defences, experimental study and challenges. Inf. Fusion **90**, 148–173 (2023)
23. M.J. Sheller, B. Edwards, G.A. Reina, J. Martin, S. Pati, A. Kotrotsou, M. Milchenko, W. Xu, D. Marcus, R.R. Colen, S. Bakas, Federated learning in medicine: facilitating multi-institutional collaborations without sharing patient data. Sci. Rep. **10**(1), 12598 (2020)
24. R.B.J. Simanjuntak, Y. Fu'adah, R. Magdalena, S. Saidah, A.B. Wiratama, I.D.W.S. Ubaidah, Cataract classification based on fundus images using convolutional neural network. JOIV: Int. J. Inf. Vis. **6**(1), 33–38 (2022)
25. W. Song, Y. Cao, Z. Qiao, Q. Wang, J.J. Yang, An improved semi-supervised learning method on cataract fundus image classification, in *2019 IEEE 43rd Annual Computer Software and Applications Conference (COMPSAC)*, vol. 2. (IEEE, 2019), pp. 362–367
26. W.G. Van Panhuis, P. Paul, C. Emerson, J. Grefenstette, R. Wilder, A.J. Herbst, D. Heymann, D.S. Burke, A systematic review of barriers to data sharing in public health. BMC Public Health **14**(1), 1–9 (2014)
27. W. Wang, X. Li, X. Qiu, X. Zhang, J. Zhao, V. Brusic, A privacy preserving framework for federated learning in smart healthcare systems. Inf. Proc. & Manag. **60**(1), 103167 (2023)
28. J. Wen, Z. Zhang, Y. Lan, Z. Cui, J. Cai, W. Zhang, A survey on federated learning: challenges and applications. Int. J. Mach. Learn. Cybern. **14**(2), 513–535 (2023)
29. I. Weni, P.E.P. Utomo, B.F. Hutabarat, M. Alfalah, Detection of cataract based on image features using convolutional neural networks. Indonesian J. Comput. Cybern. Syst. **15**(1), 75–86 (2021)
30. H. Xie, Z. Li, C. Wu, Y. Zhao, C. Lin, Z. Wang, C. Wang, Q. Gu, M. Wang, Q. Zheng, J. Jiang, Deep learning for detecting visually impaired cataracts using fundus images. Front. Cell Develop. Biol. **11** (2023)
31. J. Xu, B.S. Glicksberg, C. Su, P. Walker, J. Bian, F. Wang, Federated learning for healthcare informatics. J. Healthcare Inf. Res. **5**, 1–19 (2021)
32. S. Yadav, J.K.P.S. Yadav, Automatic cataract severity detection and grading using deep learning. J. Sensors **2023** (2023)
33. L. Zhang, J. Li, H. Han, B. Liu, J. Yang, Q. Wang, Automatic cataract detection and grading using deep convolutional neural network, in *2017 IEEE 14th International Conference on Networking, Sensing and Control (ICNSC)*. (IEEE, 2017), pp. 60–65
34. X.Q. Zhang, Y. Hu, Z.J. Xiao, J.S. Fang, R. Higashita, J. Liu, Machine learning for cataract classification/grading on ophthalmic imaging modalities: a survey. Mach. Intell. Res. **19**(3), 184–208 (2022)

Advancements and Challenges in Federated Learning for Privacy-Preserving Smart Healthcare: A Review

Habiba Akter Rimi, Md. Asaduzzaman, Md. Johir Uddin Bhuiyan, Hashibul Ahsan Shoaib, K. M. Nafiur Rahman Fuad, and Md. Anisur Rahman

Abstract Federated Learning (FL) has emerged as a revolutionary approach to machine learning, offering a significant privacy-preserving potential in data analysis in the smart healthcare sector. This systematic review focuses on comprehensively examining the advancements and related difficulties in the application of FL in privacy-preserving smart healthcare. Beginning with an overview of smart healthcare and the importance of privacy in managing sensitive health data, this review introduces FL and compares it with traditional machine learning. The methodology outlines a systematic approach including search strategies, criteria for inclusion and exclusion, data extraction, and quality assessment. FL Applications in smart healthcare, such as electronic health records, medical imaging, predictive modeling, personalized medicine, and telemedicine, have been explored. Various privacy-preserving techniques, including differential privacy, homomorphic encryption, secure multiparty computation, and data anonymization, have been discussed. This review emphasizes the advantages of Federated Learning in healthcare, emphasizing enhanced privacy of data, improved institutional collaboration, scalability, efficiency, and improved patient outcomes. It also addresses challenges, such as technical issues, privacy and security concerns, regulatory and ethical considerations, and the need for interoperability and standardization. Current trends and future directions were examined, pointing to emerging technologies, Artificial Intelligence integration, and

H. A. Rimi
Department of Computer Science and Engineering, Shanto-Mariam University of Creative Technology, Dhaka, Bangladesh

Md. Asaduzzaman · Md. J. U. Bhuiyan · K. M. N. R. Fuad (✉)
Department of Computer Science and Engineering, Bangladesh University of Business and Technology (BUBT), Dhaka, Bangladesh
e-mail: nafiurrahman373@gmail.com

H. A. Shoaib
Information Technology, St. Francis College, Brooklyn, NY, USA

Md. A. Rahman
School of Engineering and Technology, Western Illinois University, Macomb, IL, USA

M. F. Mridha and N. Dey (eds.), *Federated Learning in Health Care Technology*, Studies in Computational Intelligence 1216, https://doi.org/10.1007/978-981-96-8353-6_11

future research opportunities. The review concludes by summarizing key findings, implications for practice and policy, and recommendations for future research, providing a comprehensive understanding of FL in privacy-preserving smart healthcare and laying a foundation for further exploration in this critical field.

Keywords Federated learning · Privacy-preserving · Smart healthcare · Electronic health records · Medical imaging · Differential privacy

1 Introduction

Smart healthcare represents the convergence of advanced technologies and traditional healthcare practices to improve patient care, enhance diagnostic accuracy, and optimize treatment outcomes [1]. The integration of technologies such as artificial intelligence (AI), Internet of Things (IoT), big data analytics, and wearable devices has enabled continuous health monitoring, personalized medicine, and predictive analytics. For instance, IoT devices can track vital signs in real time, AI algorithms can analyze complex medical data to contribute to diagnosis, big data analytics can find patterns and trends that guide decisions of public health [2]. These innovations collectively contribute to a more efficient and patient-centric healthcare system.

Healthcare data is inherently sensitive, containing personal and confidential information such as patient histories, diagnostic images, genetic profiles, and treatment plans. The importance of maintaining the privacy and security of this data cannot be overstated, as breaches can direct to significant results including identity theft, loss of trust in healthcare providers, and legal penalties [3]. Furthermore, regulatory frameworks like the Health Insurance Portability and Accountability Act (HIPAA) in the United States and the General Data Protection Regulation (GDPR) in the European Union mandate strict protections for patient data [4]. Ensuring data privacy is crucial not only for compliance but also for preserving patient dignity and promoting a trustworthy healthcare environment [5].

Federated Learning (FL) is an innovative machine-learning approach that addresses the challenge of utilizing sensitive data distributed across multiple locations [6]. Traditional centralized machine-learning methods require data aggregation from various sources into a single repository, raising significant security and privacy concerns. In contrast, Federated Learning allows institutions to train machine-learning models collaboratively by sharing model parameters, not raw data. This decentralized approach enables healthcare organizations to benefit from shared knowledge while maintaining the confidentiality of their local data [7]. For example, hospitals can collaboratively improve diagnostic models without disclosing patient records, thus balancing the need for advanced analytics with privacy preservation (Fig. 1).

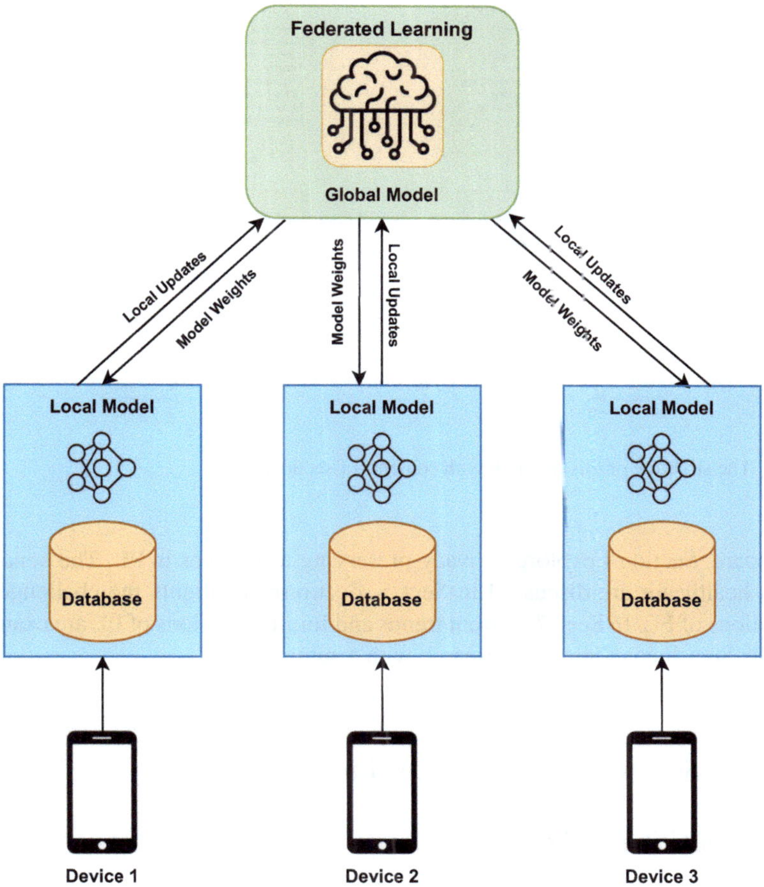

Fig. 1 Federated learning architecture

The primary objective of this standardized review is to provide a comprehensive inspection of the current state of federated learning as applied to privacy-preserving smart healthcare [8]. This review makes several key contributions to the literature (Fig. 2):

- Provides an in-depth analysis of the applications of FL in the smart healthcare context and the potential to improve healthcare services and patient care.
- Evaluates the effectiveness of privacy-preserving techniques used in FL.
- Identifies the key issues and limitations that need to be resolved to successfully implement federated learning in healthcare.
- Highlighting arising trends and outlining future research opportunities in the field.

This research is organized as follows. Section 2 provides an overview of federated learning and discusses its various types. Section 3 details FL applications in smart

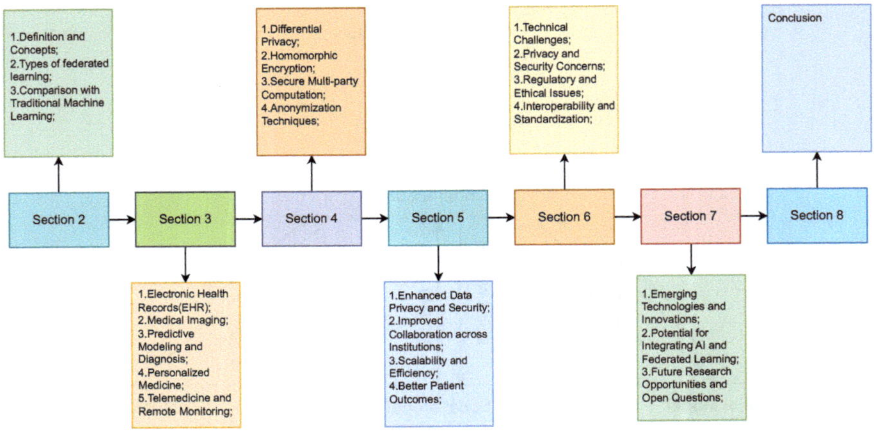

Fig. 2 The structure of this research with contents of sections

healthcare. Section 4 explores privacy-preserving techniques in FL. The benefits of FL in healthcare are discussed in Sect. 5. Section 6 highlights the challenges and limitations of FL. In Sect. 7, current trends and future directions of FL are examined. Finally, Sect. 8 concludes the research with a summary.

2 Overview of Federated Learning

2.1 Definition and Concepts

Federated Learning (FL) is an innovative machine-learning paradigm [9] designed to address the difficulties associated with data privacy and security. Unlike traditional centralized machine learning, where data are aggregated into a single repository for training models, FL allows the training of models across decentralized data sources [10]. In FL, local devices or institutions train models on their own data and share only the model updates (e.g., weights and gradients) with a central server. This central server then aggregates these updates to create a global model, which is subsequently shared with local entities [11]. The key concepts of FL include the following:

- **Decentralization:** Data remain distributed across multiple locations, reducing the risk of privacy breach.
- **Model Aggregation:** Local model updates are aggregated by a central server without accessing the raw data.

- **Privacy Preservation:** By keeping the data localized and sharing only the updates of the model, FL improves data privacy and security.
- **Scalability:** FL can be efficiently scaled across numerous devices and institutions by leveraging diverse datasets.

2.2 Types of Federated Learning

There are several categories of Federated Learning based on the data distribution and collaboration setup. The primary types are:

2.2.1 Horizontal Federated Learning

Horizontal Federated Learning, known as sample-based FL, occurs when datasets from different entities share the same feature space, but have different samples [12]. For example, multiple hospitals with similar patient attributes but different patient populations can collaboratively train a model.

2.2.2 Vertical Federated Learning

Vertical Federated Learning is known as feature-based FL, which occurs when datasets from different entities share the same sample space but have different features [13]. For instance, a hospital and pharmacy might collaborate, where the hospital has medical records and the pharmacy has purchase histories for the same set of patients.

2.2.3 Federated Transfer Learning

Federated Transfer Learning combines the aspects of both horizontal and vertical FL [14]. It is utilized when the datasets differ in sample and feature spaces both. This approach leverages transfer-learning techniques to share knowledge among different entities using distinct datasets.

2.3 Comparison with Traditional Machine Learning

There are several advantages of Federated Learning over traditional centralized machine learning, particularly in contexts where data privacy, security, and regulatory compliance are critical [15]. The key differences are as follows:

- **Data Privacy:** Data are centralized in traditional machine learning which raises the risk of privacy breaches. FL mitigates this risk by keeping the data localized and sharing model updates only.
- **Data Security:** Centralized systems are exposed to attacks owing to the aggregation of sensitive data. FL improves security by distributing the training protocol across multiple entities.
- **Regulatory Compliance:** FL helps comply with data protection regulations such as HIPAA and GDPR, which restrict data sharing across borders and institutions.
- **Scalability and Collaboration:** FL enables collaboration over multiple entities without requiring data pooling, thereby leveraging diverse datasets for robust model training.
- **Resource Utilization:** Traditional machine learning requires significant computational resources on a central server. FL distributes the load of computation across local devices and optimizes resource usage.

3 Federated Learning Applications in Smart Healthcare

3.1 Electronic Health Records (EHR)

EHR contain comprehensive patient information including medical history, diagnoses, treatment plans, medications, immunization dates, allergies, radiology images, and laboratory test results. EHRs are crucial for providing a comprehensive view of patient health; however, they also pose significant privacy risks because of the sensitive nature of the data.

Federated Learning (FL) can address these privacy concerns [16] by allowing multiple healthcare providers to collaboratively train machine-learning models on their local EHR data without sharing actual records. This collaborative approach can improve predictive analytics, patient stratification, and personalized treatment recommendations. For example, hospitals can develop better models for patient readmission prediction or identify patients at risk of developing chronic conditions by leveraging diverse data sets from multiple institutions while maintaining data privacy.

3.2 Medical Imaging

Medical imaging, including modalities such as MRIs, X-rays, CT Scans, and ultrasound, plays a vital role in treatment and diagnosis planning. However, medical imaging data are often large and sensitive, making it challenging to share them across institutions for collaborative model training.

FL enables institutions to jointly train models on distributed medical imaging data [17]. This can enhance the performance of image analysis algorithms used for applications such as tumor detection, organ segmentation, and anomaly detection [18]. By leveraging FL, healthcare providers can benefit from a larger and more diverse training dataset, resulting in more robust and accurate models. For instance, federated learning has been applied to improve the accuracy of deep learning models in detecting pneumonia [19] from chest X-rays and in identifying brain tumors from MRI scans.

3.3 Predictive Modeling and Diagnosis

Predictive modeling and diagnosis are critical components of proactive healthcare. Machine learning models can predict disease outbreaks, patient deterioration, and treatment outcomes, thus enabling timely and effective interventions [20].

FL allows the development of predictive models that utilize data gathered from multiple sources without compromising patient privacy [21]. These models can predict the onset of diseases such as diabetes, cardiovascular conditions, and infectious diseases, based on a diverse set of patient data, including demographics, clinical measurements, and lifestyle factors. For example, federated learning can be used to create models that predict heart attack risk by analyzing patient data from multiple cardiology centers, thereby enhancing the generalizability and accuracy of the model [22].

3.4 Personalized Medicine

The goal of this is to adapt the personalized medical treatment to the individual characteristics of each patient. This approach requires the analysis of huge amounts of data, including genetic information, environmental factors, and lifestyle habits.

FL can facilitate personalized medicine by enabling model development that considers diverse and geographically distributed datasets [23]. For instance, federated learning can be used to train models that predict patient responses to specific medications based on genetic profiles, thereby optimizing treatment plans [24]. By collaborating on a global scale, healthcare providers can create more effective and personalized treatment protocols, while assuring that data of the patient remains secure and private.

3.5 Telemedicine and Remote Monitoring

Telemedicine and remote-monitoring technologies are becoming increasingly important in modern healthcare, especially in providing care to patients in remote or

Table 1 A brief summary of federated learning applications in smart healthcare

Application area	Data type	FL contributions
Electronic Health Records (EHR)	Patient medical records	Privacy-preserving collaborative model training for predictive analytics, patient stratification, and personalized treatment
Medical imaging	MRIs, X-rays, CT scans, ultrasound	Improved image analysis (e.g., tumor detection, organ segmentation) by training on distributed data
Predictive modeling and diagnosis	Patient demographics, clinical measurements	Enhanced disease prediction (e.g., diabetes, cardiovascular conditions) using diverse data sources
Personalized medicine	Genetic profiles, lifestyle data	Optimized treatment plans by predicting patient-specific medication responses
Telemedicine and remote monitoring	Wearable device data, mobile health apps	Improved monitoring algorithms for chronic diseases (e.g., diabetes, hypertension) using distributed wearable data

underserved areas. These technologies rely on continuous data collection from remote-monitoring systems, wearable devices, and mobile health applications.

FL supports telemedicine by integrating data from various sources to improve remote patient monitoring and care delivery [25]. For example, federated learning can enhance algorithms that monitor chronic diseases such as diabetes and hypertension by analyzing data from wearables, such as glucose monitors and blood pressure cuffs [26]. This allows healthcare providers to offer more accurate and timely interventions based on an extensive view of a patient's health data, while preserving data privacy (Table 1).

4 Privacy-Preserving Techniques in Federated Learning

4.1 Differential Privacy

Differential Privacy (DP) is a mathematical framework used to ensure that the output of a computation does not expose too much information about any individual data point [27]. In the context of Federated Learning (FL), DP can be applied to model updates to protect the privacy of individual participant data [28]. The key concept is to attach controlled interference in the model updates before they are aggregated by the central server. This noise ensures that the contribution of any single data point is obscured, thereby preserving privacy.

The application of differential privacy in FL involves several steps:

- **Local Differential Privacy:** Noise was attached to the model updates at each client before they were sent to the central server.
- **Global Differential Privacy:** The central server aggregates noisy updates and may add additional noise to the aggregated model to guarantee privacy.
- **Privacy Budget:** A privacy budget, often denoted by ϵ, quantifies the level of privacy protection. Smaller values of ϵ provide stronger privacy guarantees, but can degrade model accuracy.

Differential privacy is widely used in FL to ensure that the training process does not compromise the privacy of individual participants, making it a cornerstone technique for privacy-preserving FL (Fig. 3).

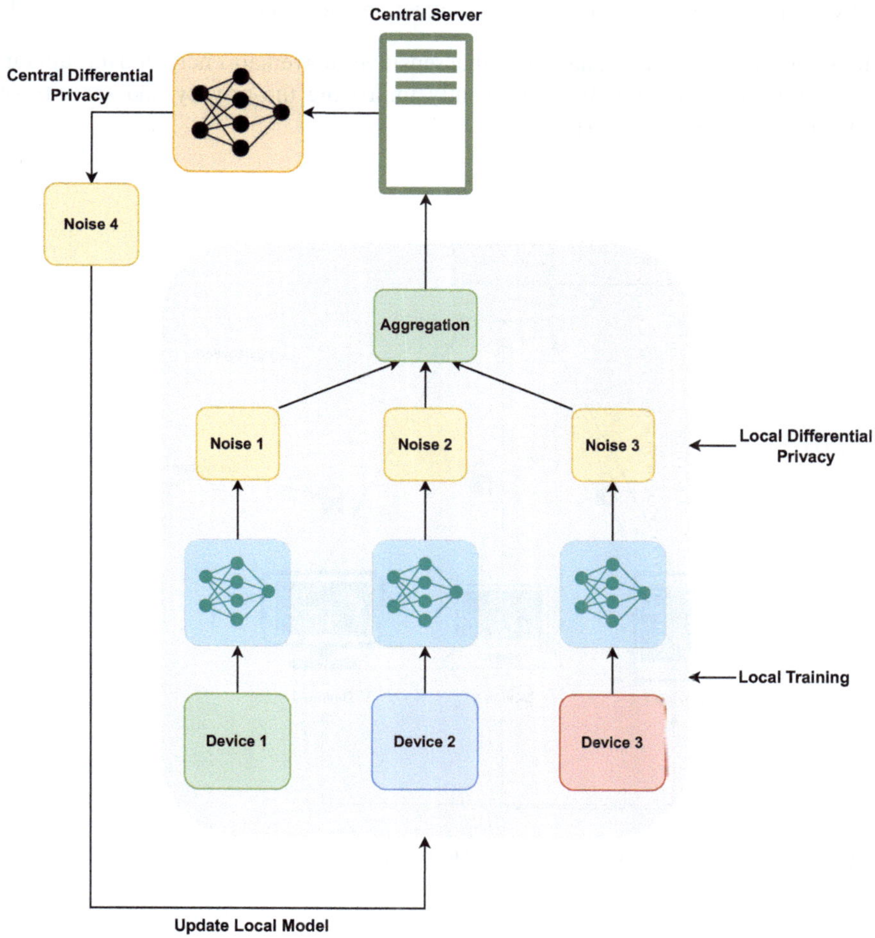

Fig. 3 Differential privacy in federated learning

4.2 Homomorphic Encryption

Homomorphic Encryption (HE) is an Encryption method that allows computations to be performed on encrypted data without decrypting it [29]. This property is particularly practical in FL, where security and privacy are paramount. HE enables the central server to aggregate encrypted model updates without requiring access to raw data [30].

The process of using homomorphic encryption in FL includes:

- **Encryption:** Each client encrypts its model updates using a homomorphic encryption scheme before transmitting them to the central server.
- **Aggregation:** The central server executes the aggregation operation on the encrypted updates.
- **Decryption:** The aggregated result is then decrypted, either by a trusted party or by using secure multi-party computation techniques, to update the global model.

Homomorphic encryption guarantees that sensitive data remain encrypted throughout the training process, thereby significantly improving the privacy and security of federated learning systems (Fig. 4).

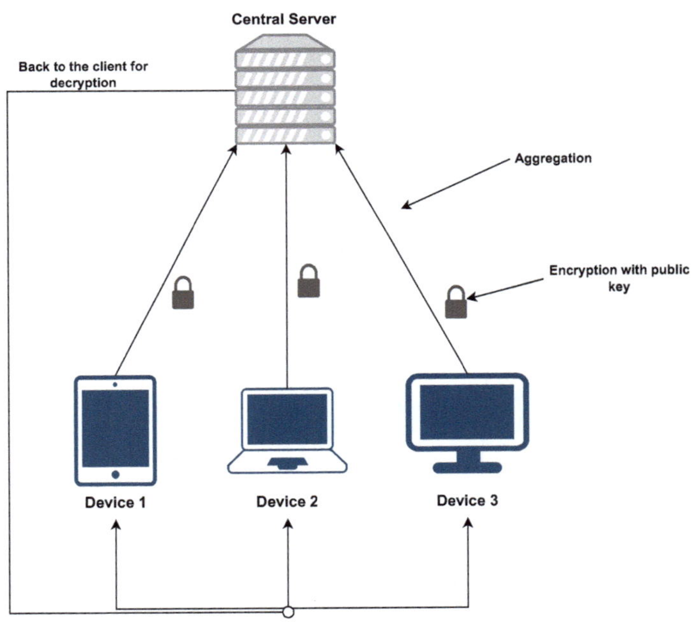

Fig. 4 Homomorphic encryption in federated learning

4.3 Secure Multi-party Computation

Secure Multi-party computation (SMPC) is a cryptographic protocol that allows multiple parties to compute a function collaboratively over their inputs while keeping those inputs private [31]. In the FL context, SMPC can be used to aggregate different client model updates securely.

The steps involved in SMPC in FL include:

- **Secret Sharing:** Each client splits model updates into multiple shares and distributes these shares to different parties.
- **Computation:** The parties perform computations on the shares without reconstructing the original updates.
- **Reconstruction:** The computational results were combined to obtain aggregated model updates.

The SMPC guarantees that no individual party has access to complete model updates, providing a high level of security and privacy. This approach is particularly useful in scenarios in which trust is distributed among multiple entities.

4.4 Anonymization Techniques

Anonymization techniques are used to remove personally identifiable information (PII) from datasets, ensuring that individuals cannot be re-identified [32]. In FL, anonymization can be applied to the data before it is used to train models as well as to model updates.

Common anonymization techniques include:

- **Data Masking:** Replacing sensitive information with generic placeholders.
- **Data generalization:** Aggregate data to a higher level of abstraction to obscure individual identities.
- **Data Suppression:** Removing specific pieces of data that could lead to re-identification.

Anonymization helps enhance privacy by ensuring that the data used in FL do not contain identifiable information, thereby reducing the threat of privacy breaches.

5 Benefits of Federated Learning in Healthcare

5.1 Enhanced Data Privacy and Security

One of the most significant benefits of Federated Learning (FL) in healthcare is its enhanced data privacy and security [33]. Traditional machine-learning models often require centralized data storage, which increases the risk of data breaches and unauthorized access [34]. By contrast, FL enables models to be trained on distributed data stored locally at different institutions. This approach ensures that sensitive patient data remain within the local environment and are not exposed to external entities [35]. By sharing only model updates and not raw data, FL reduces the attack surface and minimizes the threat of data leakage. Additionally, incorporating privacy-preserving techniques such as homomorphic encryption, differential privacy, and secure multi-party computation further strengthens data security and privacy.

5.2 Improved Collaboration across Institutions

Healthcare R&D often requires collaboration across multiple institutions to leverage diverse and comprehensive datasets. However, concerns about data privacy and proprietary information can hinder such collaboration. Federated Learning facilitates improved collaboration by enabling institutions to collaboratively train machine-learning models without the need to share raw data [36]. This decentralized approach allows each institution to preserve control over its data while contributing to the development of robust and generalizable models. As a result, FL fosters a collaborative environment in which institutions can benefit from shared knowledge and insights, thereby accelerating advancements in medical research and clinical practice.

5.3 Scalability and Efficiency

Federated Learning is inherently scalable and efficient, making it well-suited for large-scale healthcare applications [37]. In traditional centralized machine learning, aggregating and processing vast amounts of data from multiple sources can be resource-intensive and time-consuming [38]. On the other hand, FL distributes the computational workload across multiple devices and institutions, reducing the burden on any single entity. This decentralized approach not only improves computational efficiency but also allows for the integration of data from diverse sources, enhancing the model's ability to generalize across different populations and healthcare settings [39]. Furthermore, FL's scalability ensures that new data from additional institutions can be seamlessly incorporated into the training process, continuously improving model performance.

5.4 Better Patient Outcomes

Ultimately, the primary goal of implementing advanced machine-learning techniques in healthcare is to improve patient outcomes. Federated Learning contributes to this goal by enabling the development of more accurate, robust, and personalized models [40]. By using data from multiple institutions, FL helps to create models that are better able to predict patient outcomes, locate risk factors, and recommend personalized treatment plans. For example, federated models can be used to predict the onset of chronic diseases, tailor treatments based on genetic profiles, and monitor patient health in real time using wearable devices. These advancements have led to timely and effective interventions, decreased hospital readmissions, and better overall health outcomes for patients.

6 Challenges and Limitations

6.1 Technical Challenges

Federated Learning (FL) introduces several technical challenges that must be addressed to realize its full potential in healthcare [41].

6.1.1 Communication Overhead

One of the main technical challenges in FL is the significant communication overhead. Unlike traditional centralized machine learning, where all data are available in one location, FL requires constant communication between the central server and distributed clients. Each iteration of model training involves the transmission of model updates from clients to the server and the aggregated model from the server back to clients [42]. In healthcare settings, where institutions may have varying levels of connectivity and bandwidth, this communication can be slow and resource-intensive. Efficient communication protocols and strategies, such as reducing the frequency of updates or compressing model parameters, are essential to mitigate this overhead and ensure timely model convergence.

6.1.2 Model Complexity

Federated Learning models can be complex, particularly when integrating diverse datasets from various institutions. The heterogeneity of data, in terms of both quality and format, poses significant challenges for model training. Additionally, the non-IID (non-independent and identically distributed) nature of data across clients can lead to model divergence and reduced accuracy [43]. Addressing these issues requires advanced techniques in the model architecture and training algorithms that can accommodate data variability and ensure robust model performance. For instance, personalized federated learning approaches that adapt a global model to local data distributions can help achieve better model generalization and accuracy.

6.2 *Privacy and Security Concerns*

While Federated Learning aims to enhance data privacy, it is not immune to privacy and security concerns [44].

6.2.1 Inference Attacks

Although raw data are not exchanged in FL, model updates can still leak sensitive information. Malicious actors can perform inference attacks, where they analyze model updates to infer private information about the underlying data [45]. Techniques such as differential privacy can mitigate such risks by adding interference to the updates; however, this comes at the cost of model accuracy. Thus balancing utility and privacy remains a significant challenge.

6.2.2 Security Vulnerabilities

FL systems are vulnerable to different security threats, including poisoning and Byzantine attacks. In poisoning attacks, adversaries can inject malicious data or model updates to corrupt a global model. Byzantine attacks involve clients sending incorrect or misleading updates to disrupt the training process [46]. Implementing robust security measures, such as secure aggregation protocols and anomaly detection algorithms, is crucial to defend against these threats and to maintain the integrity of the federated learning process.

6.3 *Regulatory and Ethical Issues*

The implementation of Federated Learning in healthcare must navigate complex regulatory and ethical landscapes [47].

6.3.1 Data Ownership and Consent

Healthcare data are subject to strict regulations regarding ownership and consent. Institutions must ensure that they have the legal right to use patient data for federated learning and that patients provide informed consent [48]. This requires transparent policies and robust consent management systems that can handle the intricacies of data sharing and usage in a federated learning context.

6.3.2 Compliance with Data Protection Laws

FL must comply with various data protection laws such as the Health Insurance Portability and Accountability Act (HIPAA) in the United States and the General Data Protection Regulation (GDPR) in Europe [4]. These laws impose stringent requirements on data privacy and security, and non-compliance can lead to significant penalties. Ensuring that federated learning implementations meet these regulatory requirements requires comprehensive data governance frameworks and regular audits to verify compliance.

6.3.3 Ethical Considerations

There are several ethical considerations in federated learning, such as bias, fairness, and transparency. Models trained on data from multiple sources may inadvertently learn biases present in the data, leading to unfair or discriminatory outcomes [49]. Assuring that FL models are fair and impartial requires careful selection of training data and continuous monitoring of the model performance. Additionally, transparency in how models are trained and used is essential for building trust among stakeholders, healthcare providers, including patients, and regulatory bodies.

6.4 Interoperability and Standardization

Interoperability and standardization are critical for the widespread adoption of FL in healthcare.

6.4.1 Data Interoperability

Healthcare data are often stored in various formats over systems and institutions. Achieving interoperability, in which data can be seamlessly integrated and used across different platforms, is a significant challenge. Standardizing data formats and developing interoperable data exchange protocols are essential for facilitating effective federated learning [50].

6.4.2 Model Interoperability

Ensuring that federated learning models can operate across different hardware and software environments is another challenge [51]. Institutions may use different machine-learning frameworks, programming languages, and computational

Table 2 A brief summary of challenges and proposed solutions in federated learning

Category	Challenge	Proposed solutions
Technical	Communication overhead	Efficient protocols, compression, reducing updates
	Model complexity	Personalized models, robust training algorithms
Privacy & Security	Inference attacks	Differential privacy, balance privacy and utility
	Security vulnerabilities	Secure aggregation, anomaly detection
Regulatory & Ethical	Data ownership & consent	Transparent policies, consent systems
	Compliance with laws	Governance frameworks, regular audits
	Ethical bias	Bias detection, transparent model usage
Interoperability	Data interoperability	Standardizing data formats, exchange protocols
	Model interoperability	Standardized APIs and frameworks
	Protocol standardization	Industry-wide FL standards

resources. Developing standardized APIs and frameworks for federated learning can help overcome these interoperability issues and enable seamless collaboration across diverse healthcare environments.

6.4.3 Standardization of Protocols

The lack of standardized protocols for federated learning can hinder its implementation and scalability [41]. Establishing industry-wide standards for federated learning processes, including data preprocessing, model training, and update aggregation, is crucial for ensuring consistent and reliable performance. Standardization efforts can also facilitate regulatory compliance and promote trust among the stakeholders (Table 2).

7 Current Trends and Future Directions

7.1 Emerging Technologies and Innovations

The landscape of Federated Learning (FL) is rapidly evolving, driven by advancements in technology and innovative approaches to address existing challenges [52].

7.1.1 Edge Computing

Edge computing is emerging as a critical technology that complements FL by bringing computation nearer to the data source. In healthcare, edge devices such as wearables, smartphones, and local servers can perform preliminary data processing and model training, reducing latency and bandwidth usage [53]. This synergy between FL and edge computing enhances real-time data analysis and decision-making capabilities, particularly in remote patient monitoring and telemedicine applications.

7.1.2 Blockchain Technology

Blockchain technology offers a decentralized and secure framework that can enhance the trustworthiness of federated learning processes [54]. By using blockchain to manage model updates and data transactions, healthcare institutions can ensure transparency, traceability, and tamper-resistance. Smart contracts on blockchain can automate the enforcement of data-sharing agreements and privacy policies, further securing the FL ecosystem.

7.1.3 Advanced Encryption Techniques

Innovations in encryption, such as Fully Homomorphic Encryption (FHE) and Secure Multi-party Computation (SMPC), are advancing the privacy and security capabilities of FL [55]. These techniques allow computations to be executed on encrypted data without revealing the underlying information, thereby providing strong privacy guarantees. Ongoing research in optimizing these encryption methods for practical deployment in FL scenarios is a significant trend.

7.2 Potential for Integrating AI and Federated Learning

The integration of artificial intelligence (AI) with Federated Learning presents transformative potential for healthcare.

7.2.1 AI-Driven Personalization

AI algorithms can enhance the personalization of healthcare services by adapting treatments and interventions to individual patient needs. Integrating AI with FL enables the development of personalized models that learn from diverse patient data across multiple institutions without compromising privacy [56]. This integration can improve predictive accuracy and treatment efficacy, improving patient outcomes.

7.2.2 AI for Automated Model Optimization

AI techniques, such as reinforcement learning and neural architecture search [57], can be used to automate the optimization of FL models. These methods can dynamically adjust model parameters, learning rates, and aggregation strategies based on real-time feedback from the training process. This automation can enhance the efficiency and performance of FL, making it more adaptable to varying healthcare environments and data characteristics.

7.3 Future Research Opportunities and Open Questions

Despite the advancements in FL, several research opportunities and open questions remain.

7.3.1 Scalable and Efficient FL Algorithms

Developing scalable and efficient FL algorithms that can handle large-scale, heterogeneous data from numerous healthcare providers is a critical research area. Future work should focus on reducing communication overhead, improving model convergence rates, and ensuring robustness to data variability.

7.3.2 Privacy-Preserving Techniques with Minimal Accuracy Trade-Offs

While techniques like differential privacy and homomorphic encryption enhance privacy, they often come at the cost of model accuracy. There is a need for research to develop privacy-preserving methods that minimize this trade-off, ensuring that FL models maintain high accuracy while providing strong privacy guarantees.

7.3.3 Ethical and Fairness Considerations

Addressing ethical issues and ensuring fairness in FL models is an ongoing challenge. Future research should explore methods to detect and mitigate biases in federated models, ensuring that they provide equitable outcomes across different patient demographics [58]. Additionally, ethical frameworks and guidelines specific to FL in healthcare must be developed to address concerns about consent, data ownership, and accountability.

7.3.4 Interoperability and Standardization Efforts

Achieving interoperability and standardization in FL implementations is essential for widespread adoption. Future work should focus on developing standardized protocols, data formats, and APIs that facilitate seamless collaboration across diverse healthcare systems [36]. Standardization efforts can also support regulatory compliance and foster trust among stakeholders.

8 Conclusion

Federated Learning represents a revolutionary approach to harnessing the power of distributed healthcare data while maintaining stringent privacy and security standards. By enabling collaborative model training without sharing raw data, FL addresses critical challenges in data privacy, fosters improved collaboration across institutions, enhances scalability and efficiency, and ultimately contributes to improved patient outcomes. The integration of emerging technologies such as edge computing, blockchain, and advanced encryption techniques further strengthens the FL framework, while the combination of AI and FL holds promise for personalized medicine and automated model optimization. However, to thoroughly discover the potential of FL in healthcare, ongoing research must address technical challenges, security and privacy concerns, regulatory and ethical issues, and the necessity for interoperability and standardization. By overcoming these hurdles, the healthcare industry can advance toward a more secure, efficient, and patient-centric future.

References

1. M. Badawy, Integrating artificial intelligence and big data into smart healthcare systems: a comprehensive review of current practices and future directions. Artif. Intell. Evol. 133–153 (2023)
2. P. Palanisamy, A. Padmanabhan, A. Ramasamy, S. Subramaniam, Remote patient activity monitoring system by integrating iot sensors and artificial intelligence techniques. Sensors **23**(13) (2023)
3. K.M. Chong, Privacy-preserving healthcare informatics: a review. ITM Web Conf. **36**, 04005 (2021)
4. S.S. Bakare, A.O. Adeniyi, C.U. Akpuokwe, N.E. Eneh, Data privacy laws and compliance: a comparative review of the EU GDPR and USA regulations. Comput. Sci. IT Res. J. **5**(3), 528–543 (2024)
5. A.O. Adeniyi, J.O. Arowoogun, C.A. Okolo, R. Chidi, O. Babawarun, Ethical considerations in healthcare IT: a review of data privacy and patient consent issues. World J. Adv. Res. Rev. **21**(2), 1660–1668 (2024)
6. X. Yin, Y. Zhu, J. Hu, A comprehensive survey of privacy-preserving federated learning: a taxonomy, review, and future directions. ACM Comput. Surv. **54**(6) (2021)

7. M. Ali, F. Naeem, M. Tariq, G. Kaddoum, Federated learning for privacy preservation in smart healthcare systems: a comprehensive survey. IEEE J. Biomed. Health Inform. **27**(2), 778–789 (2023)
8. K. Dasaradharami Reddy, T.R. Gadekallu, A comprehensive survey on federated learning techniques for healthcare informatics. Comput. Intell. Neurosci. **2023**(1), 8393990 (2023)
9. J. Wen, Z. Zhang, Y. Lan, Z. Cui, J. Cai, W. Zhang, A survey on federated learning: challenges and applications. Int. J. Mach. Learn. Cybern. **14**(2), 513–535 (2023)
10. M. Moshawrab, M. Adda, A. Bouzouane, H. Ibrahim, A. Raad, Reviewing federated machine learning and its use in diseases prediction. Sensors **23**(4) (2023)
11. C. Zhang, Y. Xie, H. Bai, B. Yu, W. Li, Y. Gao, A survey on federated learning. Knowl.-Based Syst. **216**, 106775 (2021)
12. W. Huang, T. Li, D. Wang, S. Du, J. Zhang, T. Huang, Fairness and accuracy in horizontal federated learning. Inf. Sci. **589**, 170–185 (2022)
13. Y. Liu, Y. Kang, T. Zou, Y. Pu, Y. He, X. Ye, Y. Ouyang, Y.-Q. Zhang, Q. Yang, Vertical federated learning: concepts, advances, and challenges. IEEE Trans. Knowl. Data Eng. **36**(7), 3615–3634 (2024)
14. W. Guo, F. Zhuang, X. Zhang, Y. Tong, J. Dong, A comprehensive survey of federated transfer learning: challenges, methods and applications (2024). arXiv:2403.01387
15. S. Banabilah, M. Aloqaily, E. Alsayed, N. Malik, Y. Jararweh, Federated learning review: fundamentals, enabling technologies, and future applications. Inf. Proc. & Manag. **59**(6), 103061 (2022)
16. T.K. Dang, X. Lan, J. Weng, M. Feng, Federated learning for electronic health records. ACM Trans. Intell. Syst. Technol. (TIST) **13**(5), 1–17 (2022)
17. H. Guan, P.-T. Yap, A. Bozoki, M. Liu, Federated learning for medical image analysis: a survey. Pattern Recog. 110424 (2024)
18. S. Nazir, M. Kaleem, Federated learning for medical image analysis with deep neural networks. Diagnostics **13**(9) (2023)
19. S.H. Khan, M.G.R. Alam, A federated learning approach to pneumonia detection, in *International Conference on Engineering and Emerging Technologies* (ICEET, 2021), pp. 1–6 (2021)
20. M.S. Ibrahim, S. Saber, Machine learning and predictive analytics: advancing disease prevention in healthcare. J. Contemp. Healthcare Anal. **7**(1), 53–71 (2023)
21. D.C. Nguyen, Q.-V. Pham, P.N. Pathirana, M. Ding, A. Seneviratne, Z. Lin, O. Dobre, W.-J. Hwang, Federated learning for smart healthcare: a survey. ACM Comput. Surv. (Csur) **55**(3), 1–37 (2022)
22. A. Linardos, K. Kushibar, S. Walsh, P. Gkontra, K. Lekadir, Federated learning for multi-center imaging diagnostics: a simulation study in cardiovascular disease. Sci. Rep. **12**(1), 3551 (2022)
23. N. Rieke, J. Hancox, W. Li, F. Milletari, H.R. Roth, S. Albarqouni, S. Bakas, M.N. Galtier, B.A. Landman, K. Maier-Hein et al., The future of digital health with federated learning. NPJ Dig. Med. **3**(1), 1–7 (2020)
24. I. Dayan, H.R. Roth, A. Zhong, A. Harouni, A. Gentili, A.Z. Abidin, A. Liu, A.B. Costa, B.J. Wood, C.-S. Tsai et al., Federated learning for predicting clinical outcomes in patients with covid-19. Nat. Med. **27**(10), 1735–1743 (2021)
25. M. Hiwale, R. Walambe, V. Potdar, K. Kotecha, A systematic review of privacy-preserving methods deployed with blockchain and federated learning for the telemedicine. Healthcare Analytics 100192 (2023)
26. T. Wang, Y. Du, Y. Gong, K.-K.R. Choo, Y. Guo, Applications of federated learning in mobile health: scoping review. J. Med. Internet Res. **25**, e43006 (2023)
27. W. Liu, Y. Zhang, H. Yang, Q. Meng, A survey on differential privacy for medical data analysis. Ann. Data Sci. **11**(2), 733–747 (2024)
28. X. Yang, W. Huang, M. Ye, Dynamic personalized federated learning with adaptive differential privacy. Adv. Neural. Inf. Process. Syst. **36**, 72181–72192 (2023)
29. I. Amorim, I. Costa, Homomorphic encryption: an analysis of its applications in searchable encryption (2023). arXiv:2306.14407

30. R. Yang, T. Zhao, F. R. Yu, M. Li, D. Zhang, X. Zhao, Blockchain-based federated learning with enhanced privacy and security using homomorphic encryption and reputation. IEEE Internet Things J. (2024)
31. W. Alghamdi, R. Salama, M. Sirija, A. R. Abbas, K. Dilnoza, Secure multi-party computation for collaborative data analysis, in *E3S Web of Conferences*, vol 399. EDP Sciences (2023), pp. 04034
32. S. Mamanazarov, De-identification and anonymization: legal and technical approaches. TSUL Legal Report Int. Electr. Sci. J. **5**(1), 25–37 (2024)
33. X. Gu, F. Sabrina, Z. Fan, S. Sohail, A review of privacy enhancement methods for federated learning in healthcare systems. Int. J. Environ. Res. Public Health **20**(15), 6539 (2023)
34. R. Gupta, S. Tanwar, S. Tyagi, N. Kumar, Machine learning models for secure data analytics: a taxonomy and threat model. Comput. Commun. **153**, 406–440 (2020)
35. E.T.M. Beltrán, M.Q. Pérez, P.M.S. Sánchez, S.L. Bernal, G. Bovet, M.G. Pérez, G.M. Pérez, A.H. Celdrán, Decentralized federated learning: fundamentals, state of the art, frameworks, trends, and challenges. IEEE Commun. Surv. & Tutor. (2023)
36. M.J. Sheller, B. Edwards, G.A. Reina, J. Martin, S. Pati, A. Kotrotsou, M. Milchenko, W. Xu, D. Marcus, R.R. Colen et al., Federated learning in medicine: facilitating multi-institutional collaborations without sharing patient data. Sci. Rep. **10**(1), 12598 (2020)
37. S. Bharati, M. Mondal, P. Podder, V. Prasath, Federated learning: applications, challenges and future directions. Int. J. Hybrid Intell. Syst. **18**(1–2), 19–35 (2022)
38. L. Alzubaidi, J. Bai, A. Al-Sabaawi, J. Santamaría, A.S. Albahri, B.S.N. Al-dabbagh, M.A. Fadhel, M. Manoufali, J. Zhang, A.H. Al-Timemy et al., A survey on deep learning tools dealing with data scarcity: definitions, challenges, solutions, tips, and applications. J. Big Data **10**(1), 46 (2023)
39. M. Aledhari, R. Razzak, R.M. Parizi, F. Saeed, Federated learning: a survey on enabling technologies, protocols, and applications. IEEE Access **8**, 140699–140725 (2020)
40. R.S. Antunes, C. André da Costa, A. Küderle, I.A. Yari, B. Eskofier, Federated learning for healthcare: Systematic review and architecture proposal. ACM Trans. Intell. Syst. Technol. (TIST) **13**(4), 1–23 (2022)
41. P.M. Mammen, Federated learning: Opportunities and challenges (2021). arXiv:2101.05428
42. E. Guerra, F. Wilhelmi, M. Miozzo, P. Dini, The cost of training machine learning models over distributed data sources. IEEE Open J. Commun. Soc. **4**, 1111–1126 (2023)
43. X. Ma, J. Zhu, Z. Lin, S. Chen, Y. Qin, A state-of-the-art survey on solving non-iid data in federated learning. Futur. Gener. Comput. Syst. **135**, 244–258 (2022)
44. N. Bouacida, P. Mohapatra, Vulnerabilities in federated learning. IEEE Access **9**, 63229–63249 (2021)
45. J.C. Zhao, S. Bagchi, S. Avestimehr, K.S. Chan, S. Chaterji, D. Dimitriadis, J. Li, N. Li, A. Nourian, H.R. Roth, Federated learning privacy: attacks, defenses, applications, and policy landscape-a survey (2024). arXiv:2405.03636
46. N. Rodríguez-Barroso, E. Martínez-Cámara, M.V. Luzón, F. Herrera, Dynamic defense against byzantine poisoning attacks in federated learning. Futur. Gener. Comput. Syst. **133**, 1–9 (2022)
47. J. Prieto, Case studies in federated learning for healthcare. Federated Learning and Privacy-Preserving in Healthcare AI (2024) 91
48. A. Brauneck, L. Schmalhorst, M.M. Kazemi Majdabadi, M. Bakhtiari, U. Völker, J. Baumbach, L. Baumbach, G. Buchholtz, Federated machine learning, privacy-enhancing technologies, and data protection laws in medical research: scoping review. J. Med. Internet Res. **25**, e41588 (2023)
49. A. Balayn, C. Lofi, G.-J. Houben, Managing bias and unfairness in data for decision support: a survey of machine learning and data engineering approaches to identify and mitigate bias and unfairness within data management and analytics systems. VLDB J. **30**(5), 739–768 (2021)
50. M.G. Crowson, D. Moukheiber, A.R. Arévalo, B.D. Lam, S. Mantena, A. Rana, D. Goss, D.W. Bates, L.A. Celi, A systematic review of federated learning applications for biomedical data. PLOS Digital Health **1**(5), e0000033 (2022)

51. L.U. Khan, W. Saad, Z. Han, E. Hossain, C.S. Hong, Federated learning for internet of things: recent advances, taxonomy, and open challenges. IEEE Commun. Surv. & Tutor. **23**(3), 1759–1799 (2021)
52. A. Rauniyar, D.H. Hagos, D. Jha, J.E. Håkegård, U. Bagci, D.B. Rawat, V. Vlassov, Federated learning for medical applications: a taxonomy, current trends, challenges, and future research directions. IEEE Internet Things J. (2023)
53. H.G. Abreha, M. Hayajneh, M.A. Serhani, Federated learning in edge computing: a systematic survey. Sensors **22**(2), 450 (2022)
54. Y. Qu, M.P. Uddin, C. Gan, Y. Xiang, L. Gao, J. Yearwood, Blockchain-enabled federated learning: a survey. ACM Comput. Surv. **55**(4), 1–35 (2022)
55. G.K. Jagarlamudi, A. Yazdinejad, R.M. Parizi, S. Pouriyeh, Exploring privacy measurement in federated learning. J. Supercomput. **80**(8), 10511–10551 (2024)
56. M. Hao, H. Li, X. Luo, G. Xu, H. Yang, S. Liu, Efficient and privacy-enhanced federated learning for industrial artificial intelligence. IEEE Trans. Industr. Inf. **16**(10), 6532–6542 (2019)
57. Y. Jaafra, J.L. Laurent, A. Deruyver, M.S. Naceur, Reinforcement learning for neural architecture search: a review. Image Vis. Comput. **89**, 57–66 (2019)
58. Y.H. Ezzeldin, S. Yan, C. He, E. Ferrara, A.S. Avestimehr, Fairfed: enabling group fairness in federated learning, in *Proceedings of the AAAI Conference on Artificial Intelligence*, vol. 37 (2023), pp. 7494–7502

MIX
Papier aus verantwortungsvollen Quellen
Paper from responsible sources
FSC® C105338

If you have any concerns about our products,
you can contact us on
ProductSafety@springernature.com

In case Publisher is established outside the EU,
the EU authorized representative is:
Springer Nature Customer Service Center GmbH
Europaplatz 3, 69115 Heidelberg, Germany

Printed by Libri Plureos GmbH
in Hamburg, Germany